10のシミュレーションで地域のポテンシャルを読み解く

都市の「見える化」でまちが変わる

URBAN VISUALIZATION CHANGES OUR TOWN

小林正美　＋　明治大学都市建築デザイン研究室
Masami Kobayashi　Meiji University Urban & Architectural Design Laboratory

CONTENTS

DESIGNER: 横山ちなみ Chinami Yokoyama　　SPECIAL THANKS: 和歌山公博 Takahiro Wakayama　　田村順子 Junko Tamura　　PRINTING: シナノ書籍印刷株式会社

序 〉 はじめに
INTRODUCTION

現在、わが国では「まちづくり」という言葉がほぼ一般に定着しているが、実はそれほど昔から使われていた言葉ではない。第二次大戦後、ベビーブーマーのためのニュータウンや集合住宅を大量に建設したが、これらは皆、建設省を中心とした国家的都市計画と予算によるものであった。しかし、東京をはじめとした大都市への人口集中が顕著になりすぎて地域の過疎化が進行したため、住民に身近な行政サービスを充実させるための「地域おこし」が1980年代から真剣に議論され始め、具体的には、1995年の「地方分権推進法」の制定により、中央から地方へのまちづくり権限の制度的移譲が始まり、地方自治体が独自に都市計画の方針を立て、予算をつけて実施するような「地方分権」の時代が事実上到来した。このことは、わが国の地方自治にとっては極めて大きなインパクトとなり、このころから各自治体において「まちづくり」という議論が始まった。さらに、2004年には新たに「景観法」も制定され、地域間競争の中で、地方の独自性をいかに発見し、育成するかが全国的な課題になっていった。そのような状況の中で新たな課題として、1）地方が主体となった都市計画や、具体的な「まちづくり」を推進できる行政スタッフや民間人材の不足、2）市民の意思を「まちづくり」政策に反映させるための合意形成の方法が不明確、3）市民の意識を啓発し、自分の住む町の状況を認識するための手立てとそのための支援技術の未発達、などが挙げられた。

私たち「明治大学都市建築デザイン研究室」は1990年代から活動を始めたが、その状況に早くから着目し、そのための技術開発を一つの軸として研究を進めた。私たちの専門は建築や都市計画系であるが、「まちづくり」という言葉には、行政、財政、福祉、交通など、さまざまな領域が含まれ、あまりに包括的すぎるので、特に「まちづくりデザイン」という表現で、景観から居住空間までの物理的な空間デザインを扱うというスタンスを取っている。

地方公共団体における「まちづくり」の現場においては、市民による合意形成が最も重要なプロセスであるため、ワークショップという市民参加型イベントを開催することが一般化していった。この時、一般市民による都市空間や街並みに対する認識に関して最も不足していたのは共通の言語であった。そのため、私たちは、ワークショップにおいて、市民と情報を共有するためのツールとして、1）模型とCCD（Charge Coupled Device）カメラ、2）フォトショップによる景観シミュレーション、3）3D画像、4）VR（ヴァーチャルリアリティ）、5）まちづくりゲーム、などさまざまなツール開発を行って、住民の合意形成を推進することを心がけてきた。この本では、その中で培われた、都市を「見える化」するためのシミュレーション技術とその応用方法を紹介する。30年間の蓄積のため、現在からみれば古臭いものもあるが、これこそが日本の「まちづくり」を支えてきた技術史の一つでもあるとも言えよう。

このような背景のもとで、以下のテーマを中心に研究を続けた。

1. 記号論や視覚心理学を基礎に、街並み景観の評価や、建物の修復の方向を探る。

2. シークエンスという視界の変化を、3Dスキャナーなどを用いて詳細に記述し、建物や景観のデザインにフィードバックする。

3. デジタル情報による「都市における回遊行動」を探る。

4. 都市における建物用途の「雑多性」を顕在化し、地域の特徴を探る。

5. GIS（地理情報システム）を用い、ブラックボックス化した都市における多様な指標（建物集積や公開空地など）を顕在化する。

これらの研究テーマは、正確さを求めた科学的論拠を組み立てることを意図しているのはもちろんであるが、精緻な正確さを極めた科学研究のためではなく、あくまでも全て実践的なまちづくり活動や都市景観の改善などの社会的還元を目的にしたものである。従って、精緻さを極める読者には、多少物足りないかもしれない。最終的には、岡山県高梁市、兵庫県姫路市、神奈川県横須賀市および川崎市、東京都世田谷区下北沢地区などで、これらの研究成果を生かすことができた。

本書の構成は以下の通りである。

一章では、まず基礎理論として、都市を読み解く「5つのキーワード」を設定し、記号論からまちづくりの方法に至るまでの考え方について解説をする。

二章では、それらのキーワードを展開した分析ツールを開発し、都市を見える化する「10個のシミュレーション」について、具体的な紹介を行う。

三章では、それまでに得られた知見を応用し、実際のフィールドで実現した事例解説を行っている。特に岡山県高梁市においては1993年から30年間ゼミナール生を引率して現地で短期ワークショップを開催し、町民と交流するとともに、まちの困りごとを空間デザインにより解決するという演習を継続して実施し、市の「まちづくりデザイン」に貢献するとともに、研究・教育における貴重なフィードバックを得ることができた。このように社会とインターラクティブに関わることは、建築都市系の研究室にとっては必須の環境であるが、それが実現できたことはこのうえなく幸せであったと思っている。

Although the term "town development" ("machidukuri") is now almost universally accepted in Japan, it has not been in use for very long. After World War II, a large number of new towns and housing complexes were built for baby boomers, all of which were the result of national urban planning and budgets led by the Ministry of Construction. However, as the concentration of the population in Tokyo and other large cities became too pronounced and depopulation of the region progressed, "regional development" to improve administrative services closer to residents began to be seriously discussed in the 1980s. Specifically, with the enactment of the Law for the Promotion of Decentralization in 1995, the institutional transfer of city planning authority from the center to the regions effectively ushered in an era of "decentralization" in which local governments were able to formulate their own urban planning policies, allocate budgets, and implement plans independently. This had a tremendous impact on Japan's local governments; around this time, "town building" began to be discussed at the municipality level. Furthermore, in 2004, a new "Landscape Law" was enacted, and in the midst of interregional competition, identifying and nurturing local uniqueness became a nationwide issue. Under these circumstances, new issues included: 1) a lack of administrative staff and private-sector personnel capable of promoting local-centered urban planning and specific "town development," 2) unclear consensus-building methods to reflect the will of citizens in "town development" policies, and 3) inadequate means and supporting technology to raise citizens' awareness and make them conscious of the situation in their own towns and cities.

The "Urban Architecture and Design Laboratory at Meiji University," which began its activities in the 1990s, focused on this situation from early on and began research with the development of technology for this purpose. Our specialty is architecture and urban planning-related fields, but the term "town development" encompasses various areas such as administration, finance, welfare, and transportation; thus, we deal with physical space design, from landscape to residential space, using the expression "town development design" in particular.

Since consensus building among citizens is the most important process in the field of "town development" in local governments, it became common practice to hold citizen participation events called workshops. At this time, the most important thing lacking from the public's perception of urban space and townscape was a common language. Therefore, we developed various tools to share information with the public at the workshop, including 1) models and CCD cameras, 2) landscape simulations using Photoshop, 3) 3D images, 4) virtual reality (VR), and 5) city planning games. This book is a collection of the tools that have been developed and used to promote consensus-building among residents. In this book, we introduce the simulation technology and its application methods for "visualizing" cities that we have developed. Although some of the techniques may seem old-fashioned from today's perspective because they have been developed over a thirty-year period, it can be said that this presents the history of technologies that have supported "urban planning" in Japan. Against this background, we continued our research focusing on the following themes.

1. Exploring the direction of evaluation of townscape and restoration of buildings on the basis of semiotics and visual psychology;

2. Describing the changes in the "sequence" of the visual field using 3D scanners and other technologies and using this data as feedback for the design of buildings and landscapes;

3. Exploring "urban circulation behavior" through the provision of digital information;

4. Revealing the "miscellaneousness" of building uses in the city and exploring the relationship with regional characteristics; and

5. Using geographic information systems (GIS).

「敷地内」の計画論から、運動論としてのまちづくりへ

From "Buildings on the Site" to Town Development as Activism

アジアの都市におけるハイブリッド的多層性の形成

　西欧では、18世紀の産業革命以降、中世的な都市国家に人口が急激に集中して逼迫した住宅需要が起きたため、大量の集合住宅の建設が求められ、それまでの組積造の建物から、床・壁・天井などの建築要素に分解して再構成されたコンクリート製の住宅が大量供給された。このことが、近代建築が生まれる大きな原因になったと言われる。一方、狭隘な都市部がスラム化し非衛生になったことから、都市の外縁部に住宅建設が始まり、スプロール現象が起きが、郊外のニュータウンを計画する新しい理論として、19世紀から20世紀にかけて、田園都市で知られるエベザワー・ハワードや近隣住区で知られるクラレンスペリーなどの都市プランナーが欧米型近代都市計画の枠組みを作った。

　一方、アジア諸国においては、農業中心とした独自の中世的都市が持続的に栄えていたが、19世紀に入ると、西欧の列強諸国が植民地政策をとり、武力を持ってこれらの国を統治することになった。それまでのアジア特有の土着風土・文化の上に、西欧の近代都市計画が上書きされて、都市景観や生活様式、食文化にまで大きな影響を与えたのである。しかし、西欧の都市文化がアジア土着の固有性を全て覆いつくしたわけではない。都市を上から俯瞰して見ればよくわかるが、表層の西欧近代都市計画の象徴とも言うべきグリッド状の都市の下層に、歴史的建物や過去の伝統的なしつらえがまだら模様に散見される。また、人の目のレベルから都市景観を水平的に見れば、現代的都市景観とともに時代を超えた多様な過去の片鱗を垣間見ることができる。これは建築家 槇文彦が示した『見えがくれする都市』の一断面を語るものである。(Fig.1)は積層された概念的なレイヤーを示し、(Fig.2)は、日々私達が経験するアイレベルの都市景観の見え方を示している。

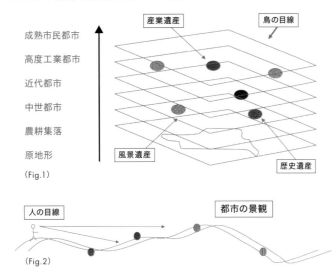

(Fig.1)

(Fig.2)

　アジアの諸都市を訪れると、香港やシンガポールでは英国的、ホーチミンやビエンチャンではフランス的、ジャカルタではオランダ的な風土や雰囲気を微妙に感じることができる。それは折衷的な建築様式や料理の文化の違いに色濃く出るが、都市景観においては、土着的建築と近代建築が混在した興味深い文化の重層性を体験することができる。植民地時代は苦労した各都市であるが、今となっては、このハイブリッド的な文化の特徴が各都市の大きな魅力の一つである。一方で、シンガポールや東京など、近代都市化が早く進んだ都市においては、世界共通の巨大資本によるグローバリズムが都心部を席巻して超高層ビルが乱立し、写真を見ただけでは、どこの都市だか判別できない傾向がますます増加している。これは、今までの土着文化に近代都市文化が重なってブレンド化されたハイブリッド文化の上に、さらに新しいグローバリズムというレイヤーが重なり始めたことを意味している。特にアジア諸都市について言えば、各都市が持つ固有の多層性を尊重し、ケビン・リンチに代表される欧米の近代都市計画理論をコピーする時代をそろそろ終わらせ、アジア独自の持続可能で環境にやさしい都市理論を私たち自身で考え出していかなくてはならない。

　アジアの一員であるわが国の首都東京や地方都市についてはどうだろうか？　徳川家康が17世紀に新たに築いた都市「江戸」は幸い帝国の植民地主義から逃れ、自ら西欧化を図り、近代都市東京へ生まれ変わった。その後、関東大震災や第二次大戦の空襲などの災害を経た後、整合的近代都市というよりは「まだら的都市構造」が形成され、個性のある「界隈」の集合として、ミクロレベル、マクロレベルでダイナミックな都市活動を続けており、典型的ハイブリッド的文化の多層性をみせている。

　一方、地方都市では、江戸時代の城下町や街道筋には歴史的資産が残されているものの、鉄道や道路などのインフラによる地域分断が激しく、未だに車に依存した生活をしている人たちが多い。少子高齢化と都市間モビリティが容易になる時代の波の中で、地方都市は人口回復について、これから激しい地域間競争にさらされる。簡単にコンパクトシティーにすべきだとは言わないが、公共交通をコアとした適度に密度がある地区を中心に、下層のレイヤーにある歴史的魅力をうまく引き出し、歩行的回遊性のある街を再構成していくことは街を活性化するための重要な戦略であり、人間中心の新しいビジョンを市民や専門家が協力して組み立てることが求められている。

都市を「見える化」する意味と、「運動論」への展開

　1980年代に建築家で東京大学教授であった原弘司は、「ものからの反撃」というエッセイの中で、ロケットを発射する際、何百という測定データを束ねて予測するよりも、1回打ち上げて、その様子を目で観察する方がひと目で状況判断がなされ、より有効であるとした。これは分析と統合というプロセスだけで物事のあり方を把握してきた現代科学に対する強いアンチテーゼであった。私たちの目は多くのことを瞬間に察知し、現状把握と次の展開への判断において、多くのデータの集積よりも重要な役割を果たしている。単なる数字の羅列ではないグラフによる視覚的表現は、要素間の関係を詳しく物語り、都市研究や合意形成には欠

かせない方法論の一つである。最近エリアマネジメントの領域で頻繁に行われている「社会実験」もトライ&エラーを許容している意味では同じ考え方に基づいていると言えよう。

　「見える化」する変換のプロセスとは、既にある多次元の事象を目で理解できる表現に翻訳する作業であり、今まで目に見えないと信じ込んでいたものを定量的に表現し、定性的な議論に持ち込むための道具である。分かりやすい例では、2次元的表現に慣れていない普通の人々にとっては、模型や三次元的スケッチは理解しやすいメディアであるため、ワークショップなどで重要な合意形成の一つのツールとなっていることが挙げられる。GISによる立体グラフのように眼に見えない都市構造を視覚化する試みも、各々の人間が勝手に思い込んでいる物事の関係や特性を異なるメディアを通して表面化することにより、関係者間の共通言語に還元し、合意形成の形成に役立っている。

　従来の建築や都市計画の世界では、さまざまな領域の関係性を定量的根拠をもとに分析し、新しい枠組みや制度を組み立てる「計画論」、具体的な物理的空間としてのスペースとそれを取り囲む「もの」の関係を考える「空間論」が主流であった。しかし、人間の日々のライフスタイルが変化し、家族像やジェンダーが変わりつつある現代において、このようなスタティックな枠組みは既に機能しない。他人同士が共生し、新たなコミュニティ意識を育む社会関係資本を生み出すべき現代においては、「運動論」というダイナミックなシステムを伴走させることが求められる。エリアマネジメントの世界で議論されている「つくる」から「つかう」への移行は、まさにこの運動論的な考えを具体的に示したものであるし、そもそも「まちづくり」という言葉には、多くの人たちを巻き込んだ運動体を形成し、合意形成しながら、自分たちの街を改良し活性化するイメージが包含されていた。その意味では、明治時代に造家学科として設立された我が国の大学における建築教育が、150年経った今も建築基準法や建築士制度の制約を受け、「敷地」内に堅固で安全な建物を建てることを中心に展開されていることは大きな問題である。60年代の寺山修司による『書を捨てよ、町に出よう』は当時の「運動論」や大学紛争のアクティビストたちに刺激を与えた名フレーズであった。社会へのプロテストが熱かった時代に戻るべきだとは言わないが、現在社会においては、大学で建築および都市を学んだ者は、「まちづくり」の一翼を担う専門家として間違いなく社会から求められているのだから、本人もマルチプレーヤーとしての自覚を持つべきであるし、加えて大学における都市建築系教育も今後ますます社会との接点を増やし、運動論的資質のある視野の広い人材を育成していく必要があるだろう。いずれにしても大学における都市建築系教育は、いまや大きく見直す時期に来ている。

Deciphering the Culture of the Multilayered City

In Western Europe, after the Industrial Revolution of the eighteenth century, the population rapidly concentrated in medieval city-states, creating an urgent demand for housing. demand led to the development and mass supply of concrete housing, which was developed through the assembly of architectural elements such as floors, walls, and ceilings ,. This was a major influence in the birth of modern architecture. Meanwhile, as dense urban areas became unsanitary slums, and housing construction began on the urban fringe, leading to the sprawl phenomenon. At this time, as a theory for planning suburban new towns, urban planners such as Evenezer Howard, known for the "Garden City", and Clarence Perry, known for the "Neighborhood Unit", created the framework for modern urban planning during the nineteenth and twentieth centuries.

In the nineteenth century, however, Western powers adopted colonial policies and ruled their colonies with military force. Modern Western urban planning overrode the indigenous climate and culture unique to Asia up to that time, and this had a major impact on urban landscapes, lifestyles, and even food culture. However, Western urban culture has not overtaken all of the indigenous Asian specificity. As can be clearly seen from a bird's-eye view of the city , historical buildings and traditional decorations of the past can still be seen in a pattern beneath the grid-like cityscape that is the symbol of modern Western urban planning on the surface. Looking at the urban landscape from the level of the human eye, we can glimpse various traces of the past that transcend time. This speaks to one element of architect Fumihiko Maki's "City with Hidden Past": Fig. 1 shows the stacked conceptual layers, while Fig. 2 illustrates how we see the urban landscape at the eye level that we experience every day. This multilayered nature of culture, especially in Asian countries, is an important point in urban studies and urban design.

What about Tokyo, the capital of our country, a member of Asia, and its regional cities? Edo, the newly established city in the 17th century by Tokugawa Ieyasu, fortunately escaped the colonialism of the empire, and was transformed into the modern city of Tokyo by its own westernization efforts. After the disasters of the Great Kanto Earthquake and the air raids of World War II, the city has developed a "fragmented urban structure" rather than a consistent modern city, and continues to exhibit dynamic urban activities at the micro and macro levels as a collection of unique "neighborhoods," showing the multilayered nature of a typical hybrid culture. On the other hand, in local cities, historical assets remain from the castle towns and roadways of the Edo period in regional cities, many inhabitants still live car-dependent lifestyles due to severe regional fragmentation caused by railroads, roads, and other infrastructure. In the coming wave of a declining birthrate, aging population, and easier inter-city mobility, local cities are facing radical inter-regional competition. Strategically drawing out the historical attractions of the lower layers and reconfiguring the city with pedestrian circulation in the central area around the public transportation hub is an important recipe for revitalizing the city.

The Meaning of Urban Visualization and its Development into Activism

Hiroshi Hara, professor at the University of Tokyo, advocated in the 1980s the importance of "looking" rather than accumulating data This discourse was a strong antithesis to modern science, which values a view of the bare essentials through analysis and synthesis alone. Visual representations by means of a graph, rather than a list of numbers, tells in detail the relationship between elements, and is one of the indispensable methodologies for research and consensus building in architectural and urban systems.

"Visualization" is a tool for quantitatively expressing and generating qualitative discussion around factors that are invisible to the eye. Sometimes, analog, easy-to-understand models and hand-drawn sketches are more easily shared with non-specialists than elaborate digital images, as they are more familiar and imaginative. Three-dimensional graphs, such as GIS graphs, are also effective tools for stimulating qualitative discussions by visualizing trends and structures of a city that are invisible. This is because the relationships and characteristics that each person assumes on his or her own can be expressed graphically through different media, reducing them to a common language among the people involved. This is undoubtedly an important tool for consensus building.

In the conventional world of architectural and urban planning, "planning theory," and "spatial theory," were the mainstream, and by mastering these theories, one could qualify as an architect in society. However, in today's world in which daily lifestyles, the image of the family, and gender are changing, such static frameworks have already lost their functionality. In addition, we are entering an era in which we must simultaneously consider a dynamic system of activism," in which strangers live together and develop a sense of community.

Those who have studied architecture and urban design at universities will undoubtedly be welcomed by society as professionals who can play a role in "town planning." Therefore, it will be necessary for urban architecture education at universities to increase its contact with society and develop human resources with activism abilities and a broad perspective. In any case, it is time for university education in urban architecture to undergo a major overhaul.

URBAN VISUALIZATION

CHANGES OUR TOWN

BASIC THEORY

CHAPTER 1

第一章　都市を読み解くための5つのキーワード

Chapter 1
1

記号論的な考え方
1. アイデンティティーを探るための交換テスト

- ■ 記号論によるメッセージの伝達方法とは？
- ■ 図像のパターン認識はどのように行われているか？
- ■ 交換テストによる「部分」と「全体」の関係を探る

私たちは、有史以来、大自然の中に住宅をはじめとした人工物を作り続けてきたが、その途中のデザインプロセスにおいても、構築物が完成した後においても、他者とのコミュニケーションは絶えず行われている。それは、言語あるいは実際の構築物を通して発せられたメッセージを媒体としたものである。ここでは、そのメッセージが伝わるメカニズムを記号論の考え方を通して考える。

まず、記号論的考え方のなかで最も重要な記号と意味の関係について考えることから始め、建築のもつ全体と部分の関係について展開する。

「意味」の構造

アメリカ合衆国の哲学者・記号論研究者であるウィリアム・モリス（1903〜1979年）は、意味論（Semantics）を「記号の意味と、意味が依存している対象の行動とを研究する」ことだと定義している。通常、曖昧に把握され、しかも広範に使用されている「意味」という言葉を明確に定義するのは非常に困難である。フランスの哲学者・記号学者であるロラン・バルト（1915〜1980年）は、『記号学の原理』の中で、「所記（意味）とは、記号の使用者がそれによって了解する『何か』なのである」と述べているが、これは明確な定義というより、逆に意味の領域の複雑さを物語っているといえよう。しかし、本論が「意味」の領域に少しでも関わっている限り、ここである程度、基本的な認識を述べておかなければならない。

意味発生のダイアグラム

一般に、情報理論や記号学などで行われる意味発生のダイアグラムは、（Fig.1）のように示される。
メッセージ：伝えるべき内容
サイン：コードによりメッセージと置きかえられたもの
コード：サインがあるメッセージを示す様定められたさまざまな約束事

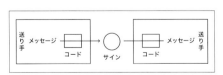

（Fig.1）メッセージ・サイン・コードの関係
送り手は自分のコードでサインを送り、受け手は
自分のコードでメッセージを受け取る

これらの関係の最も簡明な例は、暗号と暗号表の関係であろう。このダイアグラムにおいて最終的に受け手に辿り着いたメッセージが「意味」である。

建築や絵画などにおいて、このダイアグラムがそのまま有効であるとするのは問題であるが、この関係をあくまでモデルとして、いくつかの曖昧な概念を明確にできる可能性は存在している。

コードの任意性

建築や絵画などから私たちが受け取る印象・評価・連想などの総体を「意味」とした場合、論理的言語や通信業務などと異なり、その「意味」は相当に複雑である。これらを複雑にしている要因は次の2つが考えられる。
1) 送り手のコードと受け手のコードが完全に一致していないこと
2) サインがコードにより「意味」を伝えるという関係自体が、また、1つのサインになり、新たな「意味」を産み出し得るということ

1) は、コードの任意性の問題であり、2) は意味の重層性の問題である。いずれの要因にしろ、建築や絵画においては、明確なコードが設定されていない以上、すべての「意味」は、受け手の自由なコード（体験や記憶などによって積み重ねられたもの）の設定如何に関わっているといえよう。

ある建築空間が体験された時に語られる「温かみがある」とか「古典的だ」「なつかしい」といった印象や評価は、すべて個人的コードを経過した「意味」であり、普遍的な共通基盤を持ち合わせたものではない。これをここでは「心的意味」と呼ぶことにする。

また、個人間のレヴェルに留まらず、ある特定の時代や集団に共通なコードが存在する場合も一般に考えられる。それらはたとえば流行・様式・時代精神などの文化的コードの名で語られるものであるが、ここにおいても、コードは、時間・空間的に普遍性を持ち得ず、変動的であることを認めざるを得ないが、こうした時代や集団から与えられた「意味」を「文化的意味」と呼ぶことにする。「意味」がコードを拠り所とする以上、コードが個人間や時代間、集団間において多様であるという認識は、即ち「意味」そのものの多様性を示しており、「意味論」追求の大きな壁となっていることは間違いない。

「現象」領域の設定

しかし、「意味」の領域が完全にブラックボックスで、手がかりが全くないというわけではない。受け手のコードが任意的であると知りながら、送り手になおもサインを送らせ続ける力は、そうした受け手のコードの中のすべてが任意なのではなく、何か個人や時代・集団に捉われない人間共通の普遍的な部分の存在を期待しているからだと考えられる。ゲシュタルト心理学（ドイツで発祥した心理学の考え方、視覚心理も扱う）などの知覚の領域の存在は、まさに、そうした共通の部分の存在を示している。それらは、先ほどの個人的印象とは異なり、たとえば、「対比的だ」とか「閉鎖的だ」「軸性がある」といった評価として語られる

が、これらは、個人の経験や集団の好みとは独立した、人間なら誰しも共通に感ずることが可能であるような領域である。これらは「心的意味」や「文化的意味」が加えられる以前の知覚や体験の領域であり、この共通的「意味」の領域を、ここでは「現象」の領域と呼ぶことにする。この領域の設定により、漠然とした「意味」の領域は少し、その姿を表してきたといえる。即ち、(Fig.2)に示すように、「意味」の発生には、二段階のレヴェルが存在する。先ず受け手は、各人間に共通な「現象」のレヴェルにおいて、サインを知覚・体験し、次いで知覚・体験されたサインは、各人間間で多様な「心的」「文化的」レヴェルにおいて、複雑な「意味」を与えられるというメカニズムである。

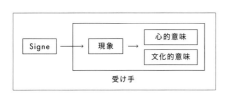

(Fig.2) サインと現象の関係
受け手は、サインをまず現象レベルで把握し後に主観的な意味に変換する

これは、即ち、コードの中に共通部分と任意的部分という2つが存在するという認識であるともいえよう。私たちが「心的」「文化的」意味について明確に語ることはほぼ不可能に近い。しかし、その「心的」「文化的」意味が拠り所とするところの「現象」については、十分語り得ると思われ、それは結果的に「心的」、「文化的」意味の領域を切り開く鍵となり得るかもしれない。

以上のような認識をもとにし、ここでは、人間による「心理」、「文化的」意味の与えられる以前の領域としての「現象」を考察の対象とする。

図像論・空間論における「現象」

図像論とは図面のような二次元的な画像認識に関する議論であり、空間論は三次元的に人間が体験する認識に関する議論のことである。図像論的に見た「現象」とは、図像のもつ特徴などにより、結果として生ずる性格を指している。一般に、図像の特徴は「パターン」と呼ばれる可視的配列の形を取り、それを把握することは「パターン認識」と呼ばれている。最終的に現われるある図像の性格を「図像特性」と呼ぶことにする。

パターン認識の概念は、顔の表情や漢字の把握、数字の自動読み取り装置の原理などによく現われている。文字という記号の解読に本質的に関わっていることからも明白なように、あるパターンは他のパターンとはっきり異なっているという条件において、自らの「図像特性」をもつことができるのである。これは基本的には、ある記号が他の記号と異なるという「差異」の構造がパターン認識にも存在していることを示している。しかし、「図像パターン」は言葉のように要素が連続的に変化するとは限らないので、言語などとは基本的に異なる点に注意しなければならない。

ロラン・バルトは「意味とは何よりも先ず『切り分ける事』だ」とし、分節 (articulation) の本質性について言及している。「パターン」と「図像特性」においては、これらは一対一対応の関係にあり、私たち人間の眼は、「パターン」の差を、「特性」の差が明解な時にはじめて認めると考えてよいであろう。従って「パターン」を変えれば「図像特性」は異なり、「図像特性」が異なっているという知覚は、「パターン」の変化を同時に示すという裏表の関係にあるといえる(Fig.3)。

具体的に、現象としての「図像特性」は軸性やリズム性、対比性等の

デザイン言語のような形で現れると考えられるが、ここで重要なことは、私たちはAという図像のもつ「特性」の内容については何も語れず、Aという図像の「特性」とBという図像の「特性」が異なっていることにしか言及で

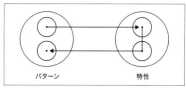

(Fig.3) 図像のパターンと特性の関係
図像のパターンが変化したときに、人間はパターン認識により、図像特性の変化を把握する

きない。それは、「特性」も「パターン」と同様、差異の構造を有しているからに他ならない。

空間論的に見た「現象」も図像論とほぼ同等であるとしてよいであろう。即ち、空間のもつ特徴は「空間パターン」であり、空間のもつ性格は「空間特性」という認識である。

例えて言えば、ある部屋の中が赤い空間パターンと青い空間パターンでは、人間が感じる空間特性は当然異なるということである。

交換テストの基本概念

交換テストとは、言語学や記号学の分野で、主に要素の分割を目的として行われる操作概念である。もともとは、音素・意味素・記号単位などの最小単位の存在を調べる目的で使用されるが、原理的には、要素と全体との関連性の度合いを調べることにより、要素を評定するものである。

ロラン・バルトは、交換テストについて、「外形(つまり能記)の面に人為的にある変化を加え、その変化が内容(所記)の面に相関的な変更をもたらすかどうかを見ることである」と述べている。これは言い換えれば、要素の交換により、どれだけ意味が変化するか(意味内容には関係しない)を調べることにより、二要素間の意味作用における優劣を決定することができることを示している。即ち、ある全体Aの中にaとbという要素があった時とき、aを同定列のa'と交換したときのAの性格の変化の度合いと、bを同系列のb'と交換したときのAの性格の変化の度合いにより、aとbの意味作用における優劣を決定できるという考え方である。この概念は、(Fig.4)の様に示される。

(Fig.4) 交換テストの概念図
全体Aの中の「b」という要素を「b'」という要素に変えたときのA全体の意味変化と、「a」という要素を「a'」という要素に変えたときのA全体の意味変化を比較し、変化の振れ幅が大きい要素の方が全体Aの中での優位に立っている。(ここでは、「a」の方がAの中で重要な役割を果たしている。)

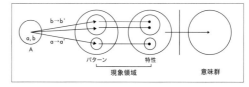

要素aをa'に変えた時の方が、要素bをb'に変えた時よりも全体の特徴の振れ幅が大きいので、要素aの方が全体に及ぼす影響力は大きい。

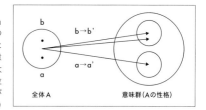

(Fig.5) パターン・図像特性・意味群の関係を示した概念図
現象領域を考えることで、誰でも共通に認識する特性を議論することができる

実際、デザインされたものの要素を調べるとき、この方法は分かりやすいが、そのまま適用するわけにはいかない。なぜなら、前述したように、意味群における変化や差を直接知る方法を私たちはもたないから

である。従って、可視的配列としての「パターン」と、その現象的「性格」としての「特性」を考えることにより、次のような（Fig.5）が成り立つ。

私たちは「特性」の差異については認識可能である。言い換えれば、それは「パターン」の差異にも関連しているので、最終的にその変化の度合いにより、a, b両要素の優劣評定は可能と考えられる。

交換の方法

交換の方法としては、次の2つが考えられる。
1）その要素の有無により、全体の「特性」の差異を調べる
2）同系列の他の要素と交換により、全体の「特性」の差異を調べる

2）の方法は、例たとえば、（Fig.5）の顔の表情の分析においては、「眼」は、ほかの要素である「鼻」あるいは「キリン」などの無関連な要素と交換してはならず、同系列の笑っているような眼、怒っているような眼と交換しなくてはいけない。これは、ロラン・バルトによる「統合」と「体系」での「体系」に当たる系列を示す。しかし、ここでは、平面の形状・スケール比という二次元しか扱っていないため、厳密な「体系」の定義は必要ないであろう。同じ幾何学形態群の中だけで交換を行えば問題はないわけである。上の二種の交換を、時に応じて使い分けることとする。

（Fig.5）人間の感情表現における各要素の機能優劣を考える

顔の表情では、交換テストにより、鼻や耳ではなく、目と眉毛が最も全体の特性を伝えていることが明白である。

本書における基本的な考えとして、地域の街並みのアイデンティティーを探るときなどに、この「交換テスト」を用いることにする。

CHAPTER 1, 1-1 ┊ SEMIOTIC APPROACH
Exchange Test to Explore Identity

- How do semiotics convey the message?

- How is the pattern recognition of iconography done?

- Exploring the relationship between "part" and "whole" through exchange testing.

Since ancient times, we have been building our houses and other artifacts in great nature, and communication with others has always taken place through the medium of language or messages sent out through artifacts. In this study, we consider the mechanism by which these messages were transmitted using the semiotic approach. We begin by examining the relationship between signs and meaning, which is the most important aspect of semiotic thinking. Then, we analyze the relationship between the whole and its parts in architecture.

Exchange Test

Basic Concepts

The exchange test is an operational concept in linguistics and semiology, primarily aimed at segmenting elements. Originally used to examine the existence of the smallest units such as phonemes, semantic elements, and symbolic units, the test is, in principle, a grading of elements by examining the degree of association between the element and the whole.

Roland Barthes describes the exchange test as "making certain artificial changes to the external (i.e., noumenal) aspect and seeing whether these changes produce correlative changes to the content (i.e., nominal) aspect." This indicates that one can determine the superiority or inferiority of the semantic action between two elements by examining how much the meaning changes (not the semantic content) as a result of the exchange of elements. In other words, if there are two elements—"a" and "b"—in the whole "A," the degree of change in the character of "A" when "a" is exchanged with "a" in the same sequence, and the degree of change in the character of "A" when "b" is exchanged with "b'" in the same sequence, the superiority or inferiority of "a" and "b" in the semantic action can be determined (Fig.4).

This method is easy to understand when examining the elements of a designed object, but it cannot be applied as is. This is because, as mentioned above, we do not have a way to directly know the changes or differences in the semantic group. Therefore, by considering "pattern" as a visible arrangement and "characteristic" as its phenomenal "character," the following Fig.5 is possible.

We can recognize differences in "characteristics." In other words, since it is related to the difference in "pattern," it is possible to evaluate the superiority of both "a" and "b" elements according to the degree of change in the final result.

Methods of exchange

There are two possible methods of exchange:

(1) The difference in the overall "characteristics" is examined by the presence or absence of the element; and

(2) Exchange with other elements of the same series to examine the difference in overall "characteristics.

In method (2), for example, in the analysis of facial expressions in Fig.6, "eyes" must not be exchanged with other irrelevant elements such as the "nose" or "giraffe," but with laughing eyes or angry eyes in the same series. This indicates a series that corresponds to Roland Barthes' "system." However, since we are dealing here with only two dimensions, the shape/scale ratio of the plane, a strict definition of a "system" is not necessary. There is no problem if the exchanges are made only within the same geometric form group. We use the above two types of exchange depending on the occasion.
In facial expressions, the exchange test makes it clear that the eyes and eyebrows, not the nose or ears, convey the most overall characteristics. The basic idea in this document is to use this "exchange test" when exploring the identity of a local townscape, for example.

Fig.1　Relationships among message, sign, and code
Fig.2　Relationship between signs and phenomena
Fig.3　Relationship between iconographic patterns and characteristics
Fig.4　Conceptual diagram of the exchange test
Fig.5　Conceptual diagram showing the relationships among patterns, iconic properties, and semantic groups
Fig.6　Functional superiority or inferiority of each element in human emotional expression

Chapter 1 / 1 記号論的な考え方

2. 差異面という指標

> ■ さまざまなものの「差異」に着目して
> 　日ごろ私たちはさまざまな認識をしている
>
> ■ 空間の「差異」の度合いをグラフに示せると
> 　さまざまなことが見えてくる
>
> ■「差異面」は建築の空間分析や都市分析に使える
> 　ビジュアルな指標である

差異の構造

1)「質」の差異と関係性

　私たちは、日ごろから多くの自然物や人工物に囲まれて生きている。特に建築を考える時、私たちは、極めて多くの「もの」を扱っていることに気づく。それは、テクスチャーであり、種々の構造材料などである。こうした雑多な「もの」を私たちはいかに群として把握し、分類的に認識できるのだろうか。それは私たちが「もの」の「質」を何か仮想し、何かの評価次元により、それが同種か異種かという判断によって識別するからに他ならない。即ち、私たちは仮想した「質」の差異の判断により、連続的な「もの」に自ら分節を与えているのである。この時、私たちは、その「もの」の「質」の内容については言及できない。なぜなら「もの」は無限の評価次元を有するからであり、「質」は私たちによって仮想されたもの、即ち、「意味」の性格を帯びているからである。私たちが唯一言及可能なのは、ある2つの「もの」の「質」が、ある次元でどれ位違っているかということのみなのである。

　「もの」に限らず、一般の事項に議論を拡張しても同様である。「質」の判断が、事項の認識の原型である。今、仮に、差異を基準にした体系として、ある建築体における形態・スケール・材料・諸機能などの多次元の事項の諸関係を扱うことが可能だとする。その時、私たちに言及不能なこれらさまざまな事項の「質」の内容から離れ、比較的少数の関係性のレベルに議論を移すことが可能となろう。もしかしたら、これが混沌とした事項間の関係に整理を与え、結果として、その建築の「質」の構造を明らかにしてくれるかもしれない。言語学・記号学の分野では既に、差異の体系として、多くの事が語られ得ると証明されている。スイスの言語学者フェルディナン・ド・ソシュール（1857～1913年）は、「記号の最も正確な特色は、他の記号でないものであるということだ」と述べている。

　実際、本来は多様で混沌とした人間の音声が差異により明確に分節され、数種の弁別的特徴の関係として体系化されて、日常会話の中で使われている事実は、関係性を扱う際の有効性を示すものである。私たちが形態を扱い、その関係性について述べるとき、「質」の評価次元を数次元に限定するならば、差異はから何かが分かるに違いない。

事項の「質」を捉える方法としての二項対立は、評価次元に差異の極端な二極を設定し、事項の「質」の度合いを評定することに他ならない。基本的には、やはり2つの事項の「質」の差異から出発した考えかたである。

2) 従来の定性分析の方法との比較

　（Fig.1）は、従来の定性分析の方法をダイヤグラム化して示している。集合した要素の性質をさまざまな角度から認識したい場合、多くの評価次元により塗り分けた地図を何枚も重ねて見ることによって、相関関係を調べるという方法がよくとられる。この方法の場合、近隣関係すなわち位置関係は、あまり問題にされず、評価軸における量的な関係のみが問題にされることが多い。

　位置関係が自由に変更可能ということは、空間を対象とする観点からすると、方法があまり有効でないということを示している。

　（Fig.2）の場合、太い線が差異の大きい関係を示している。差異の内容はどの関係においても異なっているが、差異のない部分は同質として結合することが可能である。（Fig.3）は、関係を示している。

（Fig.3）で同質の要素群が島のようになる部分と差異度の高い関係が連続する部分があことが、このモデルでの重要な部分である。この時、この島の部分を私たちは他とは異なる1つの「まとまり」として知覚しているのである。

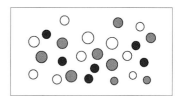

（Fig.1）従来の定性分析の方法を示した図
個々の要素の「質」の量は見えるが
要素間の関係性は把握できない

（Fig.2）差異を線の太さで示した図
個々の要素の「質」の差が明確になり
その要素が同質であるか異質であるか
把握しやすい

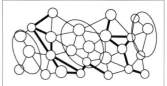

（Fig.3）同質の要素群による「まとまり」を
示した図
差異面により各要素群のの谷と山が
明確になり、「まとまり」が知覚されやすくなる

3)「差異面」グラフの考え方

　（Fig.4）は、「差異面」グラフの考え方を示した概念図である。領域Aの「質」と領域Bの「質」が異なっている場合、この境界にあるルールに従った高さの面を立ち上げる。この場合、「質」には、大きさ、形、色彩などさまざまな評価次元が含まれる。

　次に、この「差異」の考えを具体的な隣接平面の関係性において考えてみる。具体的には、一つのモデルとして、形態比率の異なる矩形平

面の「差異」をグラフ化して立ち上げ、図面上での視覚的な「差異」の知覚、および空間体験的な「差異」の知覚と対応関係になるように設定する。まず、隣接二平面の奥行き方向と接面方向のスケール比を中心にグラフを立ち上げ、プロポーションの違う空間の境界

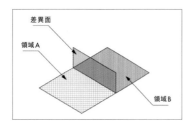

(Fig.4)「差異面」の概念図
双方の「質」の差異を境界部分の高さで表

を通過したときに人間が感じる心理的ギャップの度合いをモデルとしてグラフ化することから始める。

ここでは、人間は差異面がゼロの境界を移動した時に連続性を感じ、差異面が大きい境界を越えた時に空間の強い「結界」を感じる、ということを想定している。(本来であれば、垂直方向の空間の大きさも評価できれば、さらに体験的「差異」は精度が上がるが、ここでは平面レベルにとどめる。)

スケール比の計算方法は、どちらかの形態をもとに、奥行き方向の比と接面方向の比を計算し、その1より大きい方の値をとることにする。即ち、(Fig.5)において、Aから計測した場合、
奥行き方向のスケール比=3
接面方向のスケール比=1/2
となるが、この場合3と2というスケール比の値を採用する。

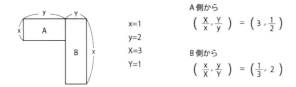

(Fig.5) 差異度の計算方法
接する(内包する)2矩形平面の奥行方向と接面方向のスケール比を計算し「差異要」の限拠とする

4)「差異面」の高さ ── 差異度

以上の規定により、プロットされた点を反対側の頂点とした平面を想定し、その面積の値の対数値を差異度とする。
今、面積をS、差異度hとしたとき、次の関係が成立する。

$$h = 10 \log S$$
$$= 10\,(\,|\log X/x| + |\log Y/y|\,)$$

h:接する(内包する)二つの矩形平面の「差異度」と定義する

たとえば(Fig.5)の場合、10(log3+log2)=7.78となる

ここでは、隣接する2要素の"質"の差異を数学的な理論の導入によってその接線に「差異面」を立ち上げ、2要素の間の関係性を視覚化するが、その高さ自体はあまり重要ではない。重要なのは集合体における「差異面」の分布状態なのである。例えば、一様に低い「差異面」の分布の中で際立って高い「差異面」を有する箇所は周囲から際立って見える箇所であり、他から強く分節されている箇所である。これはゲシュタルト心理学における"図"と"地"の明瞭に分離した箇所を指している。また、周囲より高い「差異面」が連結して面をつくっている箇所は、構成要素の軸性を読み取る手がかりとなる。(Fig.6)は2平面が包含関係になった状態を示しているが、2平面間のプロポーションによって異な

る差異面が立つことを示している。(Fig.7)は異なった平面群における差異面の分布状態を示している。

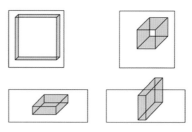

(Fig.6) さまざまな差異面の立ち方
プロポーションの異なる平面の境界には高い「差異面」が立ち、プロポーションが類似した
平面の境界には低い「差異面」が立つ
同じ形態の平面でも接する方向が90度異なると「差異面」の高さも異なる

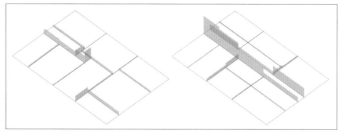

(Fig.7) 視覚化された差異面の分布状態
左図は均質な平面に異質な箇所が挿入された状態を示し、「図」的存在を示している
右図は一方向に「差異面」が立ち、他とは異なる軸線の強い空間の存在を示している

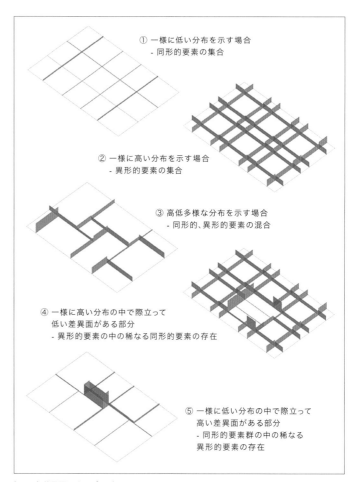

① 一様に低い分布を示す場合
  - 同形的要素の集合

② 一様に高い分布を示す場合
  - 異形的要素の集合

③ 高低多様な分布を示す場合
  - 同形的、異形的要素の混合

④ 一様に高い分布の中で際立って低い差異面がある部分
  - 異形的要素の中の稀なる同形的要素の存在

⑤ 一様に低い分布の中で際立って高い差異面がある部分
  - 同形的要素群の中の稀なる異形的要素の存在

(Fig.8) 差異面のタイポロジー
もともと均質的空間であったフィールドに人間が手を加え、異質な空間を作り出して来た
行為自体が、建築や都市形成の歴史であったことが分かる

（Fig.8）は「差異面」の分布によるさまざまなタイポロジーの在り方を示している。これらのパターンを見ても分かるように、私たちの建築空間の生成プロセスは均質な空間から、いかに差異を紡ぎだし、唯一無二の空間構成を生み出す努力であったことに、改めて気づかされる。いわゆるデザイン行為とは「差異化」をもとにした創造行為であるともいえよう。

5）建築空間および都市空間の分析ツールとしての「差異面」

（Fig.9）はル・コルビュジェのサヴォワ邸の平面図に「差異面」グラフを立ちあげた分析事例である。フランク・ロイド・ライトの住宅と比較すると、空間のメリハリが良く出ていることが分かる。このように、建築作家の作品分析や都市空間の変容などを探るときにも応用できるツールとして考えられる。

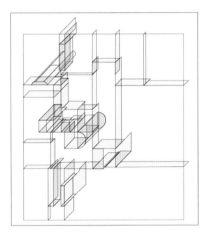

（Fig.9）「差異面」によるサヴォワ邸の空間分析事例（幾何学部分は別指標による）
ル・コルビュジェの建築作品では、均質的な空間配置に異質的な空間単位がちりばめられていることが分かる

CHAPTER 1, 1-2 ⦚ SEMIOTIC APPROACH

Basic Theory on the Concept of Differential Plane

- We recognize various things by paying attention to the "differences" between them.
- If we can show the degree of "difference" in a graph, we can see various things.
- The "differential plane" is a visual indicator to show architectural spatial analysis and urban analysis.

Structure of Difference

Differences and relationships among "qualities"

When we create an architectural environment, we deal with many "things," such as structural materials, textures, and colors. How can we grasp such miscellaneous "things" as a group and recognize them categorically? It is only because we hypothesize some "quality" of "things" and differentiate them according to certain evaluation dimensions. In other words, we segment a continuous "thing" by judging the difference in "quality." We cannot refer to the content of the "quality" of the "thing" itself because the "thing" has infinite evaluation dimensions of evaluation. This "quality" has the character of something hypothetical to us, that is, "meaning." The only thing we can judge is the relative qualities of two "things"；how they differ in a certain dimension.

Comparison with the qualitative analysis conventional method
Fig.1 shows a diagrammatic representation of the qualitative analysis conventional method.

To identify the properties of a set of elements from different angles, the neighborhood relationship is often ignored, and only the quantitative relationship in the evaluation axes is considered. This may not be effective for the space operation since the positional relationships can be freely changed. In Fig.2, the thick lines indicate relationships with large differences, and Fig.3 shows = that the island of highly differentiated relationships is the most outstanding part of the model. In this case, we perceive this island as a cohesive entity that is different from others.

Concept of "Differential Plane" Graph

Fig.4 shows a conceptual diagram of the "Differential Plane" graph concept. When the "quality" (i.e. evaluation dimensions such as size, shape, and color) of area A is different from that of area B, a plane of height according to a certain rule is set up at this boundary.

Let us consider this idea of "difference" between adjacent planes. We begin by graphing the degree of the psychological gap that humans perceive when passing through a boundary of two spaces, which have different proportions, by setting up a graph around the scale ratio between the depth and tangent directions of the two adjacent planes as a model.

The assumption here is that humans feel a sense of continuity when they move across a boundary with a zero-differential plane and a strong "boundary" of space when they cross a boundary with a large differential plane. Therefore, the scale ratio is calculated using the ratio of the depth direction and the tangent direction based on either form and taking the value that is greater than 1.

Fig.5 shows the calculation process.

When the measurement is taken from A,
Scale ratio in depth direction = 3
Scale ratio in tangent direction = 1/2
In this case, the scale ratio values of 3 and 2 are used.

Height of differential plane--degree of difference

Based on the above rules, we assume a plane with the plotted point as the opposite vertex, and the logarithm of the value of the area is the degree of difference for our optical perception.
When the area is "S" and the degree of difference is "h," the following relationship is established.

$$h = 10 \log S$$
$$= 10 \left(|\log X/x| + |\log Y/y| \right)$$

h represents the "degree of difference" between two rectangular planes that touch (encompass) each other.
For example, in the case of Fig.5, 10 (log3 + log2) = 7.78

Fig.8 shows various typologies based on the distribution of "differential planes." As these patterns show, the process of creating architectural space is an effort to weave differences from a homogeneous space to create a unique spatial configuration from scratch.

"Differential Plane" as a tool for analyzing architectural and urban space

Fig.9 is an example of an analysis in which a "differential plane" graph is created on the plan of Le Corbusier's Villa Savoi. Compared to Frank Lloyd Wright's residence, we can see that the spatial characteristics are well-defined. Thus, it can be regarded as a tool that can be applied when analyzing the works of architects and exploring the transformation of urban spaces.

Fig.1　Illustration of the conventional method of qualitative analysis
Fig.2　Differences are indicated by the thickness of the lines
Fig.3　Figure showing the "cohesiveness" of homogeneous elements
Fig.4　Conceptual diagram of a "differential plane"
Fig.5　Method of calculating the degree of difference
Fig.6　Various standing differential planes
　　　A high "plane of difference" stands at the boundary of planes of different proportions, and a low "plane of difference" stands at the boundary of planes of similar proportions. The height of the "difference plane" also differs when the planes of the same form are 90 degrees apart in the direction of contact.
Fig.7　Distribution state of visualized difference planes
　　　The left figure shows a homogeneous plane with a heterogeneous area inserted, indicating the existence of a "figure." The right figure shows the existence of a space with a "difference plane" standing in one direction and a strong axis different from the others.
Fig.8　Typology of Differential Planes
Fig.9　A case study of spatial analysis of the Savoy house by "Differential Planes　"

Chapter 1 / 2 視覚とシークエンス
1. 視覚による認知

- 視覚心理学が私たちの空間認知に大きく影響している
- 景観としての「まとまり」が地域の特徴に大きな影響を与えている
- さまざまな視覚心理学の法則にも「差異」が影響を与えている

視覚による知覚

人間の視覚とはどのような方法、認識によって情報を認知するのか。そして形態の知覚はいつ成立するのであろうか。暗い場所では視覚による知覚が不可能なように、人間の視覚が形態を知覚するためには、見るものの輝度の差異によってそのものが周囲から分化されて弁別できることが必要条件となる。そして「形態が見える」と知覚したときには、"形"と認識した領域と"その他"と認識した領域の差異の存在が把握され、「ゲシュタルト心理学」という視覚心理学における「図（figure）」と「地（ground）」という基本概念へと発展していくのである。

(Fig.1)「ルービンの盃」

この古典的な視覚プロセスの理論について見てみよう。

この「図」と「地」という性質を明確に示したものに有名な「ルービンの盃」というものがある（Fig.1）。これはデンマークの心理学者エドガー・ルービン（1886～1951年）が心理学的視点から形態視について行った研究で示した概念図であり、人間の視覚を知るうえで非常に興味深い事例である。

（Fig.1）の絵を参照すると黒い部分に注目したとき互いに向き合った二人の人物の顔が見え、白い部分に注目すると盃が見えることに気づくだろう。そしてこのとき二人の顔と盃が両方同時に見えることはなく、必ずどちらか一方が見えることに気づく。つまり、二人の顔が前面に現れたときにはそれが「図」となり盃は「地」、盃が前面に現れたときにはそれが「図」となり、二人の顔は「地」となる関係を示しているのである。また、境界線は「図」に属し「地」には属さないという特徴も、注目すべき点である。

このように人間の視覚は形態を知覚するために輝度の差異が、また情報を認知するために「図」と「地」の関係を無意識のうちに必要条件としていることがわかる。こうした造形構成の基盤概念についての認識を深めることにより、このことが実は都市構造の解析においても非常に有効的に活用できる要素であると考えること

(Fig.2) ノッリの示すローマの地図

ができる。(Fig.2)はジャンバティスタ・ノッリによる「ローマの地図」である。

これは17世紀のローマの街をゲシュタルト心理学の「図」と「地」という概念を新たな視点から捉え、都市構成の表現として示したものである。このなかでノッリは建物を黒色、外部空間を白色で表現することで私的空間と公共空間をわかりやすく示し、新たな解析方法としての都市構造の視覚化を試みている。一方、建築家の芦原義信（1918～2003年）が『外部空間の設計』で述べているように、日本の都市空間のような、建物と外部空間の間に多くの残余空間が存在する場所においては、この手法により得られる情報は曖昧であり、西欧とアジアの都市構造の違いを考える意味で興味深い。

「形態を認識すること」は、人間の視覚認知の根本であるとともに、その応用の仕方によってはさまざまな情報の認知に活用できる基本概念である。そしてその概念の発展形を都市構成の考察に用いることは非常に有効性が高い。

街並み景観における視覚の役割

人間は「見える空間（景観）」を連続的な物理空間としてとらえている。そして、この性質には、対象の大きさ、対象への距離、時間、そして視野などさまざまな要因が制限を与えている。"一度に見ることのできる空間"は人それぞれのもつ条件によって多様に異なるが、視覚的、聴覚的観点から次のような研究結果が報告されている。

『視覚系の情報処理－心理学・神経科学・情報工学からのアプローチ』によれば、単に明るさを感じることのできる視野は180度、おおざっぱに物が見える視野は約60度、文字を読むなど詳細な情報を認識できる視野は約1度であると定義されている。しかし、都市解析という観点から見ると、こういった基本的な知識に加え、"距離"に関するより詳細な情報が必要となる。すなわちどこまでを近いと感じ、どこからを遠くと感じるのか、また人間の視覚的認知の限界がどこに存在するのかという問題が重要だからである。人の遠近感覚は、個人の年齢、習慣、そして街や都市の規模によって変異するが、おおむね以下のような分類を共有化する。

①すぐ近くであると視覚認識する風景…近景
②すぐ近くではないが視覚認識の許容範囲内である風景…中景
③完全に近くではないと認識するが視界に存在する風景…遠景

しかし、同じ視覚対象物であっても、これらがどこの街、都市についても当てはまるものではない。

街並みと「～らしさ」

各地域の街並みの特性をいかに読み取るかは重要な課題である。また、地域固有の街並みとその「地域らしさ」の関係を明確にすることは重要である。ここでは、まずケーススタディーとして、現代における

古い街並みの修復において、その「地域らしさ」を発見し、育成していく場面を考えてみよう。その「地域らしさ」は私たちが街並みを知覚する時にそれぞれ主観的に感じるものであって、特定の人間によって決め付けられるものではない。また、その街並み自体は恐らく「都市の遺伝子（DNA）」のようなものを継承しながら時間の流れとともに徐々に変化していくものである。しかし、私たちが現代都市の混沌とした街並みの中からそれを見つけることは極めて困難である。それゆえに過去のイメージを単に固定化してモデル化する方向ではなく、街並みが正常に知覚できる秩序ある景観の構造を明快にすることが重要である。これは、建築家の芦原義信の著書『かくれた秩序』の中で語られていることと同じ概念である。

　たいていの場合、我々は街並みについて全体を一様に把握するのではなく部分的、断片的に知覚している。街並みは一度に多様な情報を我々の視界へ発信してくるが、それらのすべてを細部にわたって一度に知覚することはできない。我々の知覚プロセスは、それらを幾つかの「まとまり」というわかりやすい要素群にまとめることによって、全体の構造を知覚している。「まとまり」とは、他の部分との差異が顕著な部分の集合であり、容易に切り取ることができる領域と言ってよい。我々は多様な要素によって構成される街並みを知覚するために、まず、「まとまり」を探し出そうとし、それらを自らの知識や記憶と結びつけることによって、はじめてそれらが保有する固有の意味や内容を引き出せるのである。このようにして街並みを明確に知覚できる時には、多くの人がそこから何か共通なものを感じることが多い。これが誰もが共通して知覚する「現象のレベル」であることはすでに一章で述べたが、この共通に感じる内容こそが、その「地域らしさ」につながるものである。これとは逆に、「まとまり」を見出せないような場合には、街並みの特徴を明確に知覚できないため、人々はそれを「混沌」と感じる。

　近代化によって、さまざまな建材が多用された現代都市の秩序を失った街並み景観は、私たちが明確に知覚できるものでなくなったため、現代人は街並みそのものに興味を失い、日本中どこでも同じ街並みになっても気が付かないような「景観オンチ」になってしまった。こういった状況のもとでは、街並みはさらに混沌化を進め、ますますその「都市らしさ」を感じなくなることは間違いない。

　この「まとまり」をつくる構成の原理については、「ゲシュタルト心理学」の分野で良く整理されており、「近接」、「類同」、「連続」、「閉合」などの法則は、街並みの構成においても応用度が高いと考えられる。

街並みの構成に見られる諸法則

1）近接の法則

　さまざまな点について同じ特徴をもっている要素がいくつも同時に存在した場合、その距離が近い要素どうしの方がまとまりやすい。これを「近接の法則」という。

　（Fig.3）を見ると、近接している円どうしが互いにまとまりあい、a、b、cというグループをつくっている。また、（Fig.4）を見ると、eというグループに対してd、fが孤立しているように感じるだろう。このように近接している要素は互いにまとまりやすいという性質をもっている。

　これを街並み景観に置き換えてみると、（Photo1）のような街並みがあげられるが、ここでは街路空間が単なる公共道路としてではなく、人々の生活の場として内的空間性を高める役割を果しているといえる。

こうした空間は生活景として生活感がにじみだす傾向があるので、近代的な道路に比較すれば生活領域が明確なため、比較的コミュニティー感覚やシビックプライド（自分が住む場所を誇る気もち）が形成されやすい雰囲気をもっている。また、（Photo2）は、建物のセットバックによって生じたすき間や不揃いな商業施設による近代的な街並みを表している。

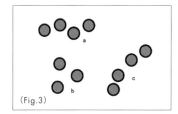
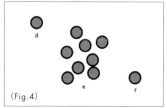

（Photo1）木造による伝統的な街並み（金沢）　　（Photo2）商業中心の街並み（原宿）

2）類同の法則

　全体を構成している色、形態、大きさ、向きなどの特徴が共通している要素は互いにまとまりやすい。これを「類同の法則」という。（Fig.5）をみると、14個の三角形のうち、3つが他と異なる色をしているが、すべてが同じ形のため、全体としては1つにまとまって見える。また、（Fig.6）をみると、14個の四角形の中に2つだけ三角形があるが、注意をあまりひかないので、同一の色をしている部分が「まとまり」として見える。全体の構成要素のなかに多少異なる性質のものが存在しても、他に何か共通の性質をもっている場合には、私たちは全体を二段階で理解し、層状の「まとまり」として捉えようとしている。

　（Photo3）はイタリアの中世都市ベネチアの街並みである。それぞれの建物は多少色の違いがあるものの、街並みを構成する建物が同じよ

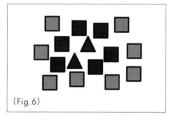

（Photo3）ベネチアの街並み（イタリア）　　（Photo4）ラストヴォの街並み（クロアチア）

うな形をしているために、全体としては一つにまとまって見える。また、（Photo4）はクロアチアのラストヴォの街並みであるが、街並みは平入り屋根や妻入り、あるいは塔が建っているなどいくつかの形の異なる建物によって構成されているが、色が同じであるために、全体として1つにまとまって見える。

この2つの例において重要なのは、異質な要素を包含し、変化に富みながらも「まとまり」のある街並みを形成しているということである。もし、これらの街並みが色も形も全く同じ建物ばかりであれば、クローンのように均質で退屈なものに違いない。一般に「まとまり」のある街並みの中に、「異質」で、きわだって見える部分がいくつか組み込まれることによって、変化に富みながらも「まとまり」のある街並みが創出されている。

3）連続の法則

同じような質をもった要素の中に同じ方向性をもった群があると、それらはまとまって見えやすい。これを「連続の法則」という。

（Fig.7）を見ると、同じ形の要素の中に同じ向きをしているa、b、c、dがまとまって見える。また、（Fig.8）のようなさまざまな形がある場合でも、その中で横に細長い要素e、f、g、h、iはまとまって見える。このように同じ方向性をもっている要素は互いにまとまりやすい性質をもっている。このような方向性は他の要素との関係性によって強められたり、あるいは弱められたりする。それ自体が強い方向性のあるものであっても、その他の要素が別の方向性をもっている場合は、かき消されてしまう。基本的には視線の動きの連続性に深く関連しており、私たちの目は常にスムーズな連続性を求めているといってよい。

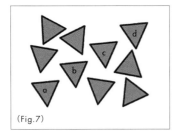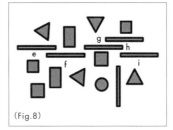
（Fig.7）　　　　　　　　　（Fig.8）

このような法則は街並みにもみられる。わが国の歴史的な街並みのいくつかに共通してみられる特徴に軒の連続性がある。また、わが国の軒に加え、欧米のコロネードという水平の方向性をもつ要素が、街並みを構成する多くの建物にみられ、それがいくつも連なることによって全体を一つにまとめている（Photo5, 6）。

（Photo5）フィレンツェの街並み（イタリア）　　（Photo6）祇園の街並み（京都）

4）閉合の法則

構成要素に共通して内に閉じ込もうとするような力が働いている場合、それらの要素は互いにまとまりやすい。これを「閉合の法則」という。

（Fig.9）のaを除く11個の円は円形で、その中心に向かって閉じ込もうとする力が働いているために、全体としてまとまって見える。それらと比較するとaの円は外側に向かう力が働いているため、まとまりにくい。また、（Fig.10）のbとcやdとeはそれぞれに二つの折れ線に、矩形を抱え込むようにして内側に閉じ込もうとする力が働いているため、互いにまとまって見える。

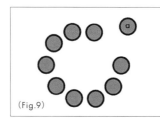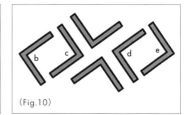
（Fig.9）　　　　　　　　　（Fig.10）

このような法則は西欧の広場を取り囲む街並みが、一つにまとまりあっているところに見られる。（Photo7）（Photo8）はイタリアのカンポ広場とナボナ広場であるが、それを取り囲んでいる街並みに、広場のカーブを抱え込むようにして内側に閉じ込もうとする力が働いているため、互いにまとまりあっている。

（Photo7）カンポ広場（シェナ）　　　（Photo8）ナボナ広場（ローマ）

4）街並み解析における「差異」の重要性

「まとまり」をつくる構成の原理である、近接、類同、連続、閉合などのいずれかの法則が各要素を結びつける働きをして「まとまり」をつくっていると考えられるが、すべてが均質ではなく、その中に「異質」を含んでいる場合は、微妙な視覚的関係性が生まれ、私たちの記憶に残りやすくなる。魅力的な街並みのイメージとしてよく挙げられる統一感、調和、リズムなどは、この「まとまり」と「異質」がどのようにバランスされるかによって、それぞれの観察者がそれぞれの受容力で感じ取るものであろう。また、ほとんどの魅力的な街並みは、これらを同時に有しているといっても過言ではない。それゆえに、複雑に絡み合っている要素間の「差異」を明確にし、街並みという事物の集合体を分かりやすく分節して把握することが街並みの構成を解き明かすことにつながると思われる。

Cognition by Vision

▦ Visual psychology has a significant impact on our spatial perception.

▦ "Cohesiveness" as a landscape has a great influence
on the characteristics of an area.

▦ Difference also influences various aspects of visual psychology.

1. Visual perception

How does human vision perceive information? When is the perception of form established? Just as visual perception is impossible in a dark place, for human vision to perceive form, it is necessary for the object to be differentiated from its surroundings by the difference in the brightness of the object. We will think about the difference between "figure" and "ground" in visual psychology, which is called Gestalt psychology. Let us look at this classical theory of visual processes.

The famous "Rubin' s cup" is a clear illustration of the nature of "figure" and "ground" (Fig.1).

Fig.1 shows that when two faces appear in front, they become the "figure," and the cup is the "ground," and when the cup appears in front, it becomes the "figure" and the two faces are the "ground." This basic concept of figurative composition shows that this factor can be used very effectively in the analysis of urban structure.

Fig.2 is a map by Giambattista Nolli, which shows the city of Rome in the 17th century as a representation of the urban configuration, based on a new perspective on the concepts of "figure" and "ground" in Gestalt psychology. In this work, Nolli attempts to visualize urban structure as a new method of analysis by representing buildings in black and external spaces in white to show private and public spaces in an easy-to-understand manner.

2. The role of vision in townscape

Humans view "visible space (landscape)" as a continuous physical space. Various factors limit this property, including the size of the object, distance to the object, time, and field of view. From the perspective of urban analysis, more detailed information on "distance" is required. In other words, it is important to know how close and far one perceives distance, and where the limits of human visual perception exist.

3. Townscape and "character"

How to read the characteristics of the townscape of each region is important. As a case study, let us first consider the situation of discovering and fostering "local character" in the restoration of an old townscape in the modern age. The "local character" is something that we all perceive subjectively when we perceive a townscape, and it is not something that can be determined by a specific person. The townscape itself changes gradually with the passage of time while inheriting the "urban gene (DNA)." However, it is extremely difficult to find it in the chaotic streets of modern cities. Therefore, it is important to clarify the structure of an ordered landscape.

Usually, we perceive townscapes in fragments. Our perceptual process perceives the overall structure by grouping them into several "cohesive" elements that are easy to understand. To perceive a townscape composed of various elements, we first try to identify "clusters," and only by connecting them with our own knowledge and memories can we draw out the unique meaning and content they possess. This common feeling of content is what leads to the "character" of the region. On the contrary, when there is no "cohesion" in a townscape, people perceive it as "chaos" because they cannot clearly perceive the characteristics of the townscape.

Modernization has resulted in a loss of order in modern cities, where various building materials have been used extensively, and the townscape is not something that we can perceive clearly. Under these circumstances, the townscape becomes even more chaotic and less and less "urban" in character.

The principles of composition that create this cohesiveness are well organized in the field of Gestalt psychology, and the laws of "proximity," "similarity," "continuity," and "closure" are considered highly applicable to the composition of townscapes.

4. Laws found in the composition of townscapes

(1) Law of Proximity

When several elements have the same characteristics in terms of various points at the same time, the elements that are closer to each other are more likely to be grouped together. This is called the "Law of Proximity." The street space in this case plays a role in enhancing the internal spatiality as a place for people to live in.

(2) Law of Similarity

Elements that share common characteristics such as color, form, size, and orientation tend to be grouped together. This is called the "Law of Similarity." Even if some of the components of the whole have slightly different characteristics, if they have some common characteristics, we understand the whole on two levels and try to see it as layered "cohesion."

Photo 3 shows a townscape of Venice, a medieval city in Italy where, despite the slight color differences, the buildings that make up the townscape have a similar shape and appear to be united as a whole. In Photo 4, the townscape of Lastovo, Croatia, is composed of several buildings with different shapes, such as those seen on roofs; they appear to be united as a whole because of their color.

(3) Law of Continuity

When a group of elements has similar qualities and the same directionality, they tend to appear to be grouped together. This is called the "law of continuity." This law can be seen in townscapes. A common feature of some of Japan' s historical townscapes is the continuity of eaves. In addition to the eaves in Japan, the horizontal directional element called a "colonnade" in Europe and the U.S. is found in many of the buildings that make up the townscape, and the many rows of the buildings unite to form a whole.

(4) Law of Closure

When there is a common force that tries to contain the elements within, the elements tend to be integrated into each other. This is known as the "Law of Closed Aggregation." This law can be seen in Western Europe, where the streets surrounding a square are all clustered into one (Photos 7 and 8).

5. Importance of "Difference" in Townscape Analysis

The principles of composition that create cohesion, such as proximity, similarity, continuity, and closure, are thought to link the various elements and create cohesion. This is why we can easily remember the cityscape as an attractive townscape. The unity, harmony, and rhythm that are often cited as images of attractive cityscapes depend on the balance between cohesiveness and heterogeneity. Most attractive townscapes have all of these elements simultaneously. Therefore, it is important to clarify the differences between the elements that are intricately intertwined and to understand the townscape as a collection of things in an easy-to-understand, segmented manner.

Fig.1   Rubin' s cup
Fig.2   Nolli' s map of Rome
Photo 1 Traditional wooden townscape (Kanazawa)
Photo 2 Commercial-oriented townscape (Harajuku)
Photo 3 Venice townscape, Italy
Photo 4 Rastovo townscape, Croatia
Photo 5 Townscape of Florence, Italy
Photo 6 Townscape of Gion, Kyoto, Japan
Photo 7 Piazza del Campo, Siena
Photo 8 Piazza Navona, Rome

2. 光源投射法による視覚

- ■ 視覚的なシークエンスが感情の変化を引き起こす
- ■ 空間を知覚する方法には「見る」と「見える」の2つがある
- ■ 「見える」は可視領域として光源投射法で「見える化」できる

「見える」空間および「見る」空間とその範囲

人間が日ごろ体験している変化する空間はさまざまな感情の変化を引き起こす。たとえば低い天井の廊下からアトリウムに出た瞬間には驚きを感じるし、細い路地を歩く時にはその先に何があるのだろうかとワクワクする感情に引き立てられる。このような空間の変化をシークエンスと呼ぶが、ここでは、視覚的なシークエンスについて考える。

人間の五感の中で、最も情報を収集する能力に長けているのは、「眼」である。眼による視覚により、人間はさまざまな環境情報を確認し、自分の行動を決めているといってよい。しかし、我々は常に眼を動かし、また体も移動しながら、周囲の情報を獲得しているのが日常である。また、無自覚に「見える」視覚と、意識をもって「見る(注視する)」視覚では、情報収集範囲や情報量が異なっていることに注意しなければならない。一般に、視覚を得ている場所を「視点場」と呼ぶが、景観などを扱う議論では、「見える」情報を、視点が動かない「静止的視点場」を基準に扱う場合が多い。一方、視点が動く場合、「見る」視覚が連続的に変化することを、「シークエンス」という。フランスの建築家 ル・コルビュジェ(1887~1965年)は、空間の中に「時間」の概念を取り込み、ランプ(傾斜路)を多用し、シークエンスの面白さを引き出し

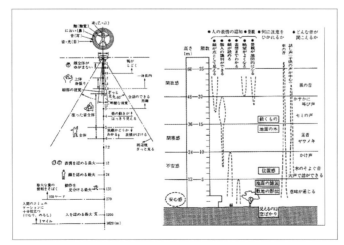

(Fig.1) 視覚・聴覚のヒューマンスケール(加藤孝義「空間のエコロジー」新曜社より)
左が水平所方向の視覚距離、右図が垂直方向の視覚距離を示している

た建築家として有名であるが、3次元的空間を4次元的空間にしたといってもよいであろう。

(Fig.1)は、加藤孝義の『空間のエコロジー』からの引用であるが、自然界における対象と人間の水平的距離および垂直的距離により、人間の視覚情報が大きく異なることが分かる。これは、広く「見える」範囲を示した考え方である。人間は動物の一種であるので、以前は敵から自

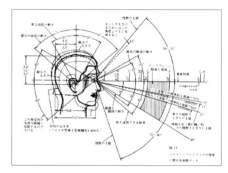

(Fig.2) リチャード・ドレヒュスによる
適正視覚の(樋口忠彦「景観の構造」より)
眼の高さより少し低めの視野範囲が
標準的であるとされている

分を守るために長距離までが視覚範囲になっていたと思われるが、近代都市においては、短くなっていると予想される。一方、(Fig.2)は人間工学に詳しい工業デザイナーであるリチャード・ドレヒュスによる視覚範囲の図である。ここでは、垂直方向の視覚範囲が示されているが、注視することによる「見る」範囲は限定されており、「見える」範囲とは

異なっていることを示している。本書ではこれらの指標を参考に、それぞれの視覚分析対象に適合した距離基準や視野角度を使用していく。

(Fig.3)は畑田豊彦による研究であるが、人間の視野範囲には、①弁別視野 ②有効視野 ③注視安定視野 ④誘導視野 ⑤補助視野があり、①~③は人が自覚して空間を認識し、「見る」範囲であり、④は無自覚に範囲内にある空間の情報を取得できる、「見える」領域であるとされてい

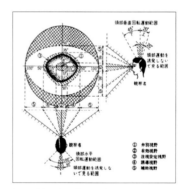

(Fig.3) 視野内の情報受容特性
(畑田豊彦の研究による)
頭部の回転による視野範囲の
広がりを示している

る。また、⑤は無自覚に空間の情報を刺激として受け取る「環境として見える」領域なので、ここでは対象にはしない。以上のことから、ここでは「見える」視野範囲の分析をする際に、④を基本的に基準とし、個別の分析対象などにより立体的な視野範囲を求めていくことにする。

光源投射法の考え方

人間の見える範囲を考えるときに、アナログ的方法であるが、「視点場に点光源があり、そこからの減衰する光が到達する範囲」を視野範囲(ビジブルエリア)としてシミュレーションする方法を「光源投射法」と名

付けて検討を行う。この「光源投射法」には2つの利用法があり、1) 人間の「見える」視野範囲を表すことができるために、移動する時に見えるミクロ的なシークエンスの検討に使えるとともに、2) ランドマークと呼ばれる、景観のシンボルや都市の特徴ある部分が、どの範囲から知覚できるかというマクロ的な視点の検討に使うことができる。

　1) の具体的な方法については、二章「光源投射法による「見える」シークエンス」で述べる。ここでは、1) と2) についての分かりやすいイメージを示しておくことにする。

二次元地図　　三次元地図　　ビジブルエリアの分析対象　　点光源による対象の照射

（Fig.4）光源投射法による「見える」視野範囲のイメージ
2D地図から3Dイメージを立ち上げ光源を移動させながら光が照射される範囲を描き計算する

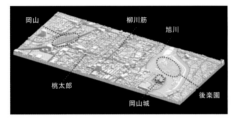

岡山　　　　柳川筋　　　旭川
桃太郎
岡山城　　　　後楽園

（Fig,5）岡山駅から岡山城方面へ向けた周辺地図

現代・過去・自然が
岡山城

（Fig,6）岡山城に光源を置いて
視野範囲を分析した地図
白い領域が岡山城を視認できる範囲であるが、ビルが立て込み、極めて視野範囲が狭くなっていることが分かる

岡山駅

（Fig,7）岡山駅に光源を置いて
視野範囲を分析した地図
白い領域が駅を視認できる範囲であるが、桃太郎大通りに強烈なビューコリドー（視線の軸）が生まれていることが分かる。逆に、岡山駅から桃太郎大通りを介して後楽園までを視認できることが分かる

スカイツリーを見ることができるエリア
スカイツリー

（Fig.8）スカイツリーを見ることができる地域
地上643mのデジタル放送用の電波塔であるので、広範囲から視認されることが絶対条件である。白い領域がその視認範囲であるが、広域にわたって殆どの領域が満たされていることが分かる

Visibility using Light Source Projection Method

▥ Visual sequencing causes emotional changes.

▥ There are two ways to perceive space: "seeing" and "looking."

▥ The "visible area" can be visualized using the light source projection method as the visible domain.

"Seeing" and "looking" space and its range

The changing space that humans experience every day induces various emotional changes. Among the five human senses, the eye is the most adept at gathering information. We realize that humans confirm various environmental information and determine their actions through visual perception. In general, the place from which certain vision is obtained is called the "viewpoint," and in discussions on landscapes, "visible" information is often treated based on a "static viewpoint" where the viewpoint does not move. However, when the viewpoint moves, the continuous change of the "seen" vision is called a "sequence." In the next section, we discuss visual sequences.

Fig.1, which draws from Takayoshi Kato's "Ecology of Space" (Shinyosha, 1986), shows that human visual information differs greatly depending on the horizontal and vertical distances between humans and objects in the natural world. This is an idea that broadly indicates the range of "seeing." Fig.2 is a diagram of visual range by Richard Drefus, an industrial designer with expertise in ergonomics. Here, the vertical visual range is shown, which indicates that the "looking" range is limited and differs from the "seeing" range. In this document, we use these indices as a reference and use distance criteria and viewing angles that are adapted to each visual analysis subject.

Fig.3 describes a study conducted by Toyohiko Hatada, which states that the range of human visual field includes (1) discriminative visual field, (2) effective visual field, (3) stable visual field for gazing, (4) guided visual field, and (5) auxiliary visual field, with (1) to (3) being the range in which people are aware of space and "see," and (4) is the "visible" area, in which one can acquire information about the space within the range unconsciously. Based on the above, when analyzing the visual field range in this study, we use (4) and seek the three-dimensional visual field range according to individual analysis targets.

Concept of Light Source Projection Method

When considering the range of human vision, the "light source projection method" is used to simulate the range of vision (visible area) as "the range where a point light source exists at the viewpoint and the attenuating light from the source reaches the viewer." This method has two applications: 1) it can be used to examine the microscopic sequences that people can see when they move, and 2) it can be used to examine the macroscopic perspective from which landmarks can be perceived. The specific method for 1) is described in Chapter 2.

Fig.1　Human scale for vision and hearing (from "Ecology of Space" by Takayoshi Kato, Shinyosha)
Fig.2　Diagram of proper vision by Richard Dreyfuss (from "Structure of Landscape" by Tadahiko Higuchi)
Fig.3　Information reception characteristics within the visual field (based on research by Toyohiko Hatada)
Fig.4　Image of the "visible" area using the light source projection method
Fig.5　Map of the area from around Okayama Station to Okayama Castle
Fig.6　Map analyzing the visible area with the light source placed at Okayama Castle
Fig.7　Map analyzing the visible area with the light source placed at Okayama Station
Fig.8　Areas where the Sky Tree can be seen

- 動線には「ツリー構造」と「セミラティス構造」の違いがある
- 「セミラティス構造」のほうが回遊性が豊かで都市に欠かせない
- 都市の回遊性には「多様な用途」「小さな街区」などが求められる

2タイプの動線の考え方

「都市はツリーではない」といったクリストファー・アレクザンダーは、パターン・ランゲージの主唱者でも知られるが、人間の移動動線の「ツリー構造」「セミラティス構造」により、明快な論理を展開した。

ツリー構造とは、(Fig.1)の左図のように、樹状に部屋が配置され、メインの動線を通らないと他の部屋にいけないリニアーな空間組織である。一方、セミラティス構造は、右図のように、連結状に部屋が配置され、人々は動線経路についていくつかのオプションをもつことができる。

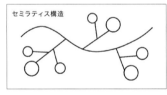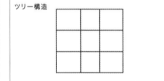

(Fig.1) ツリー構造とセミラティス構造の動線図
左図は中央のメイン動線を通らないと各部屋に入れない
右図は、扉のつけ方で、いろいろな部屋を通過しながら複数の動線を確保できる

これらの動線をベン図や動線組織図で描くと、(Fig.2)のように示される。ベン図を見ると、「ツリー構造」はオーバーラップした領域がなく、各部屋は上位の空間に包含され、そこを通過しないと他に部屋に行けない。一方、「セミラティス構造」はさまざまにオーバーラップした領域があり、どちらにも属さないような空間が複数存在する。動線組織図もそれを示しており、前者はヒエラルキーがはっきりしており、後者はヒエラルキーがあいまいなネットワーク型になっている。

建物のレベルでいえば、古典建築ではセミラティス構造が多くみられるが、近代建築では、合理性を

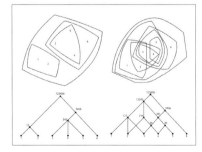

(Fig.2)ツリー構造とセミラティス構造の
ベン図と動線組織図
左図はヒエラルキーの明解な構造を示しており、
右図はネットワーク型の構造を示している。

求めるためにツリー構造が多用されてきた。このことはその後の建物再生の際に大きな差として現れ、近代建築の他用途への転換が難しいことが指摘されている。(DOCOMOMO ―近代建築の記録と保存を目的とする国際学術組織―で指定されている近代建築も解体の危機にさらされることが多い。)

クリスファー・アレグザンダーは、都市についても言及し、「都市はツリーではない」と強く主張した。ヨーロッパの中世都市における回遊動線は、まさしく彼が唱えた多様なネットワークによる歩行者動線が都市を豊かにしている(Photo1,2)。しかし、20世紀に建設されたブ

(Photo1) ブラジリア(ブラジル)

イタリアの中世都市シェナと近代都市計画によるブラジリアの比較
道路のネットワークを見てもセミラティス構造とツリー構造の違いが良くわかる

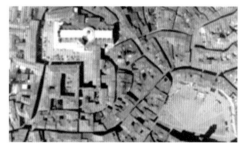

(Photo2) シェナ(イタリア)

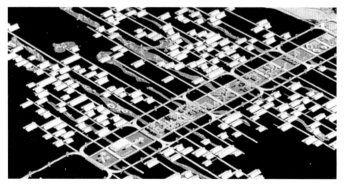

(Photo3)東京計画1960

ラジリアや丹下健三による「東京計画1964」では、強烈なツリー構造が構想され、強く批判されてきたのである（Photo3）。

一方、20世紀後半にアメリカの都市計画に影響を与えたジェーン・ジェイコブスは、1）都市には複数の主要用途があること、2）街区は小さくて街角が多くあること、3）都市にさまざまな年代の建物が混在すること、4）ある程度の人口密度があること、が豊かな都市の生活空間を生む最低条件であると主張した（Fig.3）。この考え方も、同様に都市のセミラティス構造を強調した考え方である。

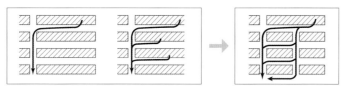

（Fig.3）ジェーン・ジェイコブスの主張による小さいブロックの必要性
街区の長さが短い方が、様々な選択肢が生まれ、回遊性につながる

そういう意味では、一本の商店街のみによる地域はリニアーな発展しか望めないが、複数の商店街が面的に広がっている地域は、回遊動線も生まれ、都市の歩行体験も豊かになることが予想される。東京の渋谷、新宿、下北沢などの繁華街はこの典型であるともいえよう（Fig.3）。また、単なる商業集積だけではなく、昨今は、デジタル情報の普及で新たな回遊動線のきっかけが生まれ、新しい都市の魅力の発見などに貢献していることも特筆すべき点である。

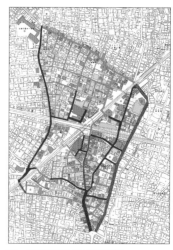

（Fig.4）下北沢地区の画的広がり

Circular Flow in the City

- There is a difference between a "tree structure" and a "semi-lattice structure" for human flow path.

- The semi-lattice structure is one of the essential structures in the city because it allows for richer circulation.

- Multiple uses and small city blocks are required for urban circulation.

2 Type of Circulation System

Christopher Alexander, who advocated pattern language, developed a clear logic of the tree structure and the semi-lattice structure of the human flow path. The tree structure is a linear spatial organization in which rooms are arranged in a tree-like pattern, as shown in the left diagram of Fig.1, and one must pass through the main flow path to reach other rooms. On the other hand, the semi-lattice structure, as shown on the right, arranges rooms in a connected manner and allows several options for flow paths.

These flow paths can be depicted in a Venn diagram or a flow-line organization diagram as shown in Fig.2. The Venn diagram shows that the tree structure has no overlapping areas, and each room is encompassed by a higher-level space with a clear hierarchy, through which no other room can be accessed. On the other hand, the semi-lattice structure has various overlapping areas, and multiple spaces do not belong to either of them where the hierarchy is ambiguous.

At the building level, semi-lattice structures are common in classical architecture, whereas tree structures have been used more frequently in modern architecture in the pursuit of

rationality. This has been noted as a major difference in the subsequent regeneration of buildings, making it difficult to convert modern architecture to other uses and putting those designated by DOCOMOMO in danger of demolition.

Christopher Alexander also referred to the city and strongly asserted that "the city is not a tree." The pedestrian paths of diverse networks of circulation lines in Europe's medieval cities enrich them, just as he advocated. However, in Brasilia, built in the 20th century, and in the "Tokyo Plan 1964" by Kenzo Tange, a strong tree structure was conceived and strongly criticized.

Jane Jacobs, who influenced American urban planning in the latter half of the 20th century, argued that the minimum conditions for a rich urban living space are: 1) a city should have multiple primary uses, 2) city blocks should be small and have many street corners, 3) the city should have a mix of buildings of various ages, and 4) the city should have a certain population density. This idea similarly emphasizes the semi-lattice nature of the city.
In this sense, an area with merely one shopping street can only expect linear development, whereas an area with multiple shopping streets spreading out over an area could create a circular flow of traffic and enrich the urban walking experience. The downtown areas of Shibuya, Shinjuku, and Shimokitazawa are types of such areas.

Fig.1 Flow path diagram of a tree structure and a semi-lattice structure
Fig.2 Venn diagram of tree and semi-lattice structures and the flow path organization diagram
The left diagram shows a structure with a clear hierarchy, and the right diagram shows a network-type structure
Photos 1 and 2 Comparison of the Italian medieval city of Siena and Brasilia with modern urban planning
Photo3 A Plan for Tokyo 1964 by Kenzo Tange
Fig.3 Jane Jacobs' argument for smaller blocks. Shorter city block lengths create various options and lead to circulation
Fig.4 Map of Shimokitazawa

Chapter 1 - 3

都市における人間の行為と多用途性
2. 変化する人間の行為と対応する空間

- ■ 常に変化する人間の行為に建築空間は対応できるか?
- ■ 建物の保存・再生がこれからは主流になる
- ■ 建物の用途規制が多用途性を求める都市活性の足かせになっている

建築における形態と機能の関係

　人間は朝から晩までさまざまな行為を行っており、常にそれに適した場所や空間を求めて動いている。都市や建築の空間は、予めこうした人間の行為を想定して計画することが常に求められてきた。しかし、時代とともに人間の行動様式が変わり、交通手段や情報収集の方法が日進月歩に変わる現在、自ずと事前に用意された都市や建物の空間が、変化する人間の行為に対応しきれない場面が起こるのは当然の状況である。具体的には、人間の行為は、時間レベル、季節レベル、年代レベルで常に変化しており、どのように、それぞれの空間を人間の行為(機能)に対応させていけばよいのかが重要な課題となっている。

　建築における「形態」と「機能」の関係は永遠の課題であり、ル・コルビュジェに代表される機能主義者たちは、「空間は機能に従う」とし、機械のように合理的な空間を良しとした。一方、形態を重視するルイス・カーンは、「形態は機能を啓発する」と述べ、自然に人間の行為が起きるような空間形態が重要だとした。ここでは、建築のレベルと都市のレベルから人間の行為と空間の関係から掘り下げてみよう。

プログラム(機能・用途):
人間の行為の内容を記述し、タイプ分類したもの
空間(スペース):
人間の行為を保証する物理的うつわ

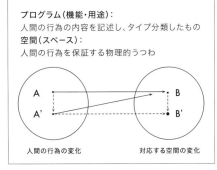

人間の行為の変化　　　　対応する空間の変化

(Fig.1)プログラムと空間の
対応関係を示したダイアグラム
人間の行為AがA'に変化した時に空間BがB'に
変化できるかどうかがポイントとなる

　(Fig.1)は、プログラム対スペースの対応関係を示しており、Aという行為がA'に変化したときに、Bという空間はB'に変化して対応し得ているのかどうかがポイントとなる。対応できていない場合は、A'のための空間が用意されていないため、人は極めて不便を強いられることになる。しかし、実際の生活では、それなりに空間側が柔軟に変化して対応しているのであまり不都合が生じていない。それでは、実際に、どのように可変的プログラムに対し空間は対応しているのか検証する。

　以下では、1)室内のレベル、2)部屋群のレベル、3)建物のレベルに分けて、それぞれの対応法を考えてみる。

1) 室内レベル
家具やパーティションによる柔軟な対応

(Photo1) ファンズワース邸

　ミース・ファン・デル・ローエの「ファンズワース邸」のような、ユニバーサル・スペースにおいては、家具でコーナーをつくり、さまざまな行為にフレキシブルに対応しており、今ではワンルームマンションなどでも応用されている。

2) 部屋群のレベル
インタラクティブスペースによる緩やかな接続

　アトリウムのような中間領域を設け、部屋群の境界を緩やかにして流動的にする。

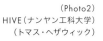

(Photo2)
HIVE(ナンヤン工科大学)
(トマス・ヘザウィック)

3) 建物のレベル
ノベーションなどによるプログラム変化への対応

　時代の要求に合わない建物の再生がよく行われる。また、長期的なスケルトンと短期的なインフィルの組み合わせによる対応。

(Photo3)
Next 21のスケルトン・インフィル

　では、都市においてはどうだろうか? 建築基準法では、フレキシブルな建物用途などは存在せず、必ず建物の用途(プログラム)を決めなくてはならない。また、その用途に従った建物の仕様、建てられる用途地域が厳密に定められている。次に挙げる「用途制限の概要」のように、さまざまなビルディングタイプが具体的な名前で指定され、都市における位置づけが決定されている。しかし、現代社会では、すでに特定の建物用途が使われなくなったり、新しい業態が生まれたり、複合的なビルディングタイプが誕生する可能性があるのに、このように行政が固定的にルール付けをしてしまったところに大きな無理が生じている。さらに言えば、都市計画図の中に定められた用途地図の色分けにより、住居地域が守られているということは理解できるが、全体に建てられるビルディングタイプが制限されてしまい、多用途的な生

活圏の創造が難しくなっていることは大きな課題である。

　建物用途については、建物内部の話にとどまらない。ジェーン・ジェイコブスは、都市には、3個以上の用途が混在していることが好ましいとした。たとえば、新宿歌舞伎町のように建物用途が多様である界隈には、さまざまな人たちが集積し、独自の地域性を醸し出されている。一方、大手町のようなオフィス街と呼ばれるところでは、同じような服装の人たちが行き来しており、ある種均質な風景を見せている。このように、建物用途の有り様は、地域がもつ個性にも強くつながっており、「建物用途」、「それを使う人々」、「地域の個性」をセットとして考えていくことが重要である。

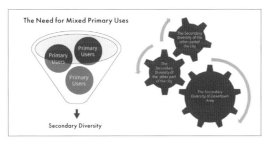

（Fig.2）ジェーン・ジェイコブスと彼女の主張

用途地域による建築物の用途制限の概要

（注1）本表は、改正後の建築基準別表第2の概要であり、すべての制限について掲載したものではない。
（注2）都市計画区域、火災地域、と畜場、火葬場、汚物処理場、ごみ焼却場等は、都市計画区域内において用途地域の指定の如何にかかわらず都市計画決定が必要など、別に規定あり。

（Fig.3）用途地域による建築物の用途制限（国土交通省）

Changing Human Actions and Corresponding Spaces

■ Can architectural space respond to ever-changing human activities?

■ Preservation and renovation of buildings will be the mainstream in the future.

■ Regulations on the use of buildings are hindering urban revitalization that seeks multi-use.

Relationship of Function and Space

Humans engage in various activities from morning till night, always moving in search of suitable places and spaces. The planning of urban and architectural spaces has always been required to respond to such human activities. However, as human behavior changes over time in relation to the evolution of transportation systems and information technologies, spaces in cities and buildings can no longer cope with the shifts. In this section, we examine the relationship between human action and space from the perspective of architecture and that of the city.

The terms are defined as follows:
Program (function/use): the content of human action which is typified
Space (space): the physical container that guarantees human action

Fig.1 shows the relationship between the program and space. The key point is whether space B can change to space B' when human action A changes to A'. If it cannot respond, people will feel inconvenienced because no space is provided for A'. Then, how does the space actually respond to the variable program?

In the following, we will classify the correspondence between 1) the room level, 2) the room group level, and 3) the building level.

1) Room level: Flexible response through furniture and partitions

In universal spaces, as in the examples of Mies van der Rohe's "Farnsworth House" and Philip Johnson's "Glass House," various areas are created with furniture to flexibly accommodate different actions. Similar gestures can be seen in studio apartments.

(2) Group of rooms: Loose connection through interactive spaces

Intermediate areas, such as atriums are created to make the boundaries among the groups of rooms more loosely connected.

(3) Building level: Renovations, etc.

A combination of long-term skeleton structures and short-term infills (interior systems) responds to the needs of building rehabilitation.

What about in cities? Under the Building Standard Law, there is no such thing as flexible building use. Therefore, the program must be determined since specifications and the zoning districts in which they can be built are strictly defined by the building use and designate the position in the city. However, in today's society, with the emergence of new businesses and complex building types, such fixed government rules are facing criticism.

Building uses are not limited to building interiors. Jane Jacobs argued that cities should have a mix of three or more uses. For example, a neighborhood such as Shinjuku Kabukicho, where buildings are used for various purposes, attracts a range of people, creating its own local character. On the other hand, in an office district such as Otemachi, people travel to and from work, which creates a kind of homogeneous landscape. Thus, it is important to consider building use, the people who use the buildings, and the character of the area as a set.

Photo 1　Glass House by Philip Johnson
Fig.1　Diagram showing the correspondence between program and space
Photo 2　Atrium at Hive by Thomas Heatherwick (Nanyang University, Singapore)
Photo 3　Next 21's skeleton-infill
Fig.2　Jane Jacobs and her insistence on the importance of versatility
Fig.3　Summary of building use restrictions by zoning district

GIS による都市の見える化
都市の情報をレイヤ化する

特別寄稿　小池博

■ まちづくりで有効なGIS（地理情報システム）の
　役割とは？

■ 地図情報とエクセル情報を連動させて
　空間情報分析を行う

■ 2D地図を目的に応じて3D化し
　都市を「見える化」する

地図のデジタル化や大量データの処理が可能なGIS

　GISとはGeographic Information Systemの頭文字をとった略称で、日本語では「地理情報システム」と訳される。GISの定義は国土交通省国土地理院によると、「地理的位置を手がかりに、位置に関する情報をもったデータ（空間データ）を総合的に管理・加工し、視覚的に表示し、高度な分析や迅速な判断を可能にする技術」と定義されている。位置に関する情報をもったデータ（空間データ）は2種類存在し、1つは地図上にプロットされた記号そのものにデータが記録されている場合と、もう1つは地図が同じ大きさの正方形グリッドで分割され、そのグリッドに情報が記録されている場合である。前者はベクターデータ、後者はラスターデータという異なったデータシステムとしてGISでは扱われている。

　GIS自体は情報を処理・管理するシステムのことを指しているが、そこで扱われるデータは原則的に地図データであるので、アウトプットとしてはなんらかの情報を包含した地図であることが多い。そのため、一般的には情報を含んだ地図そのものをGISと表現している場合が多い。ここでは、地図も含めた広義の意味でのGISを扱うものとする。

　コンピューターの性能が飛躍的に向上し、インターネットの普及に伴い、ITが日常生活に急速に広まっていった。コンピューターの性能の向上に伴い、それまでGISを操作するうえでネックとなっていた地図のデジタル化や大量データの処理が容易に行えるようになり、GISも次第に社会へ広まることとなった。日本においても、1992年に日本地理情報システム学会が発足し、GISの研究が進められたが、1995年の阪神・淡路大震災においてGISの災害時における有用性が見直され、政府もGISの研究・普及に本腰を入れるようになった。現在は、無料のGISソフトも普及し、今後はいかにそれらのデータを活用し、新たな価値を見出していくかが重要な課題である。

　広義では情報を含んだ地図そのものがGISと認識されているが、地図情報を処理・管理するだけの狭義のGISではなく、広義のGISに含まれている一般社会への汎用性を考慮していくことが、まちづくりへの応用などにおいても重要である。

情報分析ツールとして活用できるGIS

　現在、スマートフォンの普及などで、飛躍的にGISは一般生活の中で使われている。
（1）地図としての活用：
　　GPSと連動したナビゲーションシステムなど
（2）情報ストックツールとして活用：
　　紙媒体をデジタル化した行政の都市計画図など
（3）情報閲覧ツールとして活用：
　　検索エンジンによる地図閲覧など
（4）コミュニケーションツールとして活用：
　　「Pokémon GO」などの双方向的交流など
（5）情報分析ツールとして活用：
　　都市デザインやまちづくりへの有効な応用など
　特に、5）が私たち、建築やまちづくりに関わるものが空間分析を行うときには極めて重要なポイントである。まず簡単なところでは、エクセルデータとの互換性があるため、通常エクセルで行うような統計的分析も行うことができ、さらにはSPSSなどの統計解析ソフトと連動させて大容量多変量解析なども行うことが可能である。また、GISならではの分析手法としては、表データだけでなく、地図データを処理することにより空間分析が行えることが挙げられる。GISの地図上にプロットされたデータを使用し、AとBの重なる部分を抽出したり、AとBの両方の属性を有する新しいデータを作成したりすることが地図データを操作することで可能となる。これは位置情報をビジュアル情報としてストックできるGISの優位性である。また、検索機能についても、さまざまな地図データを利用して、特定の属性を持つ地域の抽出など、応用範囲は大きい。以下では、具体的な活用法の一部を紹介する。

GISデータの活用事例

　GISで扱うデータは、主に地図上にプロットされた要素（以下、フィーチャという）そのものをデータとして扱う（1）ベクターデータ、地図にグリッドで網掛けし、そのグリッドごとの情報を記録している（2）ラスターデータ、そして画像などによる（3）イメージデータに分けられる。一般的には、写真画像などのデータはピクセルで構成されており、ラスターデータの1つとして扱われることが多い。ここでは便宜上、画像データのことはイメージデータとして統一し、エクセル表データを持つグリッド上の地図情報をラスターデータと呼ぶこととする。

（1）ベクターデータ

　ベクターデータとは、数式、直線、曲線を使用し、フィーチャそのものがIDをもつデータである。イラストレーターで主に扱っているデータがベクターデータ、フォトショップで主に扱っているデータがラス

ターデータとするとわかりやすい。ベクターデータでは、同じレイヤの中でフィーチャ同士を重ねることも可能である。ベクターデータにおけるフィーチャには以下の3種類が存在する。

A) ポイントデータ

点で表現されるものがポイントデータである。厳密には地図上では位置情報しかもたず、大きさの情報はもたない。ポイントデータは大きさをもたないゆえに、純粋に位置の分布を検証したいときに利用される。面積情報を超えたスケールでの分布状況の検証にはポイントデータが極めて有効である（Fig.1）。

（Fig.1）都内の公開空地のプロット事例
ポイントデータにより正確な位置同定が可能となる

また、ポイントデータの特長として、TINデータの作成が挙げられる。TINは、簡単にいえばポイントデータの中なかで距離的に近い3つの点を結び、三角形の面を作成し、その集合体として面の3Dモデルを作成する手法である。これにより、2次元データを3次元イメージに変換することができる。

B) ラインデータ

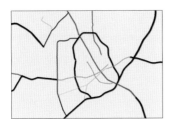

（Fig.2）太さと色分けによる道路の種類表現
地図上の線情報の内容を異なった太さや色で示してレイヤ状に重ねると多様な総合情報が得られる

ラインデータとは、文字通り、地図上で線により表現されたデータである。したがって、リニアな位置情報はもつものの、ポイントデータと同じく大きさや幅はもっていない。ただし、ポイントデータと同様に表現的に太さを変えた表現ができるため、異なった線種を特定な記号として適用することが可能である（Fig.2）。

ラインデータには、地図上の表現として大きく分けて3つの種類が考えられる。1つ目は、鉄道や道路のように実際には幅があるものの、地図情報上のラインデータとして扱われるものである。

2つ目は、崖や尾根などの実際の空間でも幅をもたない境界情報である。これはそのままラインデータをあてはめればよい場合が多く、表現に関しては地図の認識のしやすさが目安となる。3つ目は、実際の都市空間には現出してこないが地図情報上ではラインデータとして表現が適しているものである。等高線はその1つであるが、自然条件を示す情報で地理的条件を把握するうえで極めて重要である。ラインデータはその性格上、都市空間の境界や地理的な条件を表現するために利用されることが多い。

C) ポリゴンデータ

ポリゴンデータとは、角形や円などの「閉じた」図形で表現されたデータであり、いわゆる「面」のデータである。したがって、ポリゴンデータの各フィーチャは、位置のほかに面積や形状が表現されている。このポリゴンデータはラスターデータと並んで、都市空間を

分析するうえで極めて重要な表現手法となる。大きさや形状が表現されているため、建物や敷地といった一般的な都市構成要素をデータ化するのに最適である。また、表現手法としても、各フィーチャの周囲のアウトラインと内側の面とで別々の表現をすることが可能で、目的に応じさまざまな表現手法を適用できる。

ポリゴンデータは実際の都市空間に存在している造形物のありのままを図形としてデータ化しているので、アウトラインを境界線として扱うと、ラインデータで囲われた図形として認識することができる。ポリゴンデータの3D化については、フィーチャに高さを与えて、上へ引き出すことで3D化する。また、建物のポリゴンデータにGLからXmというように具体的に数値を与えて、都市内の建物群の3D表現ができる。たとえば地形の3DモデルをTINで作成した場合、そのTINを立ち上げの基準面として指定すれば、すべてのポリゴンデータがTINの面を基準に、その上に立ち上げることが可能である（Fig.3）。

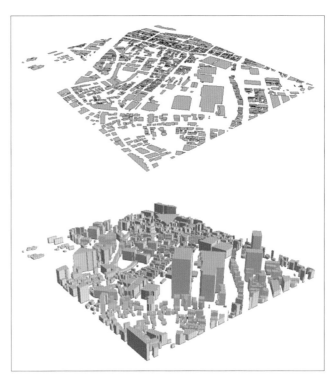

（Fig.3）ポリゴンデータによる都市の3D表現
建物のポリゴンデータに建物高さを与えると、3Dの都市表現が可能となる

このようにそれぞれのデータの特徴が異なるので、分析の目的や最終的な表現を考慮したうえで、ポイントデータ、ラインデータ、ポリゴンデータの選択が行われる。

（2）ラスターデータ

ラスターデータと呼ばれるものは、ベクターデータが要素そのものをデータとして取り扱うのに対し、地図をグリッドで分割し、そのグリッドごとにデータを記録するものである。このグリッドのことをセルと呼ぶ。

ラスターデータにはその特徴によって大きく2つに分けることができる。1つは隣り合うセルのデータ自体が連続的につながっており、便宜上その中間値や平均値などが各セルに与えられているラスター

データであり、これを「連続面ラスターデータ」と呼ぶ。もう1つのラスターデータは、各セルになんらかの数値や記号が記録として埋められたラスターデータである。したがって、隣り合うセルどうしは必ずしも連続していない。このようなあらかじめ決められたなんらかのデータがセルごとに与えられたラスターデータを「離散的ラスターデータ」と呼ばれる（Fig.4）。離散的ラスターデータについて考察を進めると、連続的なものを含め、あらゆるデータはクラス分けを行うことで、離散的ラスターデータに変換できる。すなわち、連続面ラスターデータで扱う連続したデータについても、意図的に境界線を設け、ゾーンで分割し、各ゾーンに数値や記号を割り当てることで離散的ラスターデータへ変更

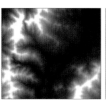

（Fig.4）連続的ラスターデータによる標高表現（左）と離散的ラスターデータによる土地利用マップ（右）
連続的なセルと離散的なセルの扱いによりデータの性格が異なり、目的に合わせて応用する

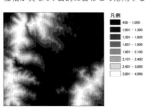

（Fig.5）連続的ラスターデータを離散的ラスターデータに変換した標高の色分け
連続的な標高表現を高さ別に分節し色分けすることで高さ関係を明確に表現することができる

できる。実はこの連続面ラスターデータから離散的ラスターデータへの変換は、ラスターデータにおけるGISのもっとも重要な解析機能のひとつとして挙げることができる（Fig.5）。

離散的ラスターデータの場合、用途地域マップなどはそうであるが、隣り合うセルで同じデータとなり、ある程度、エリア分けが可能となるケースが多い。その場合、表現としては極めてベクターデータのポリゴンデータに近い表現となる。実際、ラスターデータをベクターデータへ変換することは可能であり、ベクターデータへ変更するためには離散的ラスターデータへ変換しておく必要がある。

（3）イメージデータ

イメージデータは画像データなので、そのファイル形式は多岐にわたる。その操作性能上よく利用されるのはTIFデータであるが、これはデータ損失がほとんどなく最も正確な画像データの一1つである反面、容量が大きくなる傾向がある。都市スケールのレベルで操作する場合、TIFデータだと重すぎて効率が悪いことが多いため、圧縮し解像度を落としたJPEGファイルが都市スケールのGISでは利用されることが多い。

イメージデータを扱う時に注意すべき事項としては、ほかのデータとの位置情報の整合性である。なんらかの座標系を有している画像データであれば調整は容易であるが、そうでない場合、他のベクターデータやラスターデータの位置と画像データの位置のずれの調整が困難であることが多い。GIS以外のソフトの操作技術も必要とされるため、可能であれば位置に関する座標系を有した画像データを使用することが望まれる。

地図情報とエクセル表データ

GISはその名前が示す通り、地理的情報がベースとなる。すなわち

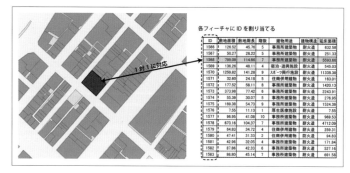

（Fig.6）1対1に対応するフィーチャとエクセル表データ（ベクターデータ）

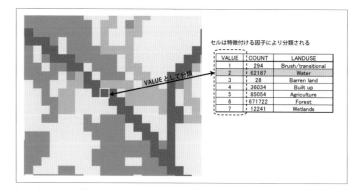

（Fig.7）1対1に対応するフィーチャとエクセル表データ（ラスターデータ）

構成要素としては地図情報と、それを補完するエクセル表データとなる。地図情報で具体的な位置や、ベクターデータであれば大きさや形状まで表現され、それぞれの地図上の要素に帰属するさまざまな情報がエクセル表データとして記録される。ここで重要なことは、ベクターデータの場合、フィーチャ（地図上の要素）に対し、エクセル表データが1対1で対応していることである。具体的に説明すると、各フィーチャにはIDが自動的（あるいは意図的も可）に割り当てられ、1つのフィーチャがエクセル表データ上で、1つの行として表わされる。その行数は記録したい情報の数により、半無限大に増やすことができる（Fig.6）。

一方、ラスターデータの場合、各セルに情報が埋め込まれているのであるが、各セルを特定する因子としての情報は各レイヤでひとつずつである。また、ベクターデータのような個々のセルに対してのIDは存在しない。したがって、エクセル表データとしては、セルの特長を特定する因子がVALUEとして表示され、その因子ごとのデータ表示となる（Fig.7）。

地図情報とエクセル表データが関係づけられているということは、すべての情報がエクセルによる統計分析が可能ということである。地図情報の中でIDづけられているフィーチャがエクセルにより統計分析可能であるため、その結果を即時にフィーチャに反映することができる。そして、新たに統計分析で作成されたデータを具体的に色分け表現することも容易に行える。このように、地図情報とエクセル表データが関係づけられていると、統計分析による空間評価が容易に行えるため、調査成果の伝達力やデザインコンセプトの説得力を強化するうえで、GISは極めて有効なツールといえる。

実際のGISによる分析作業においては、いくつもの条件をもったファイルを何枚も重ね、空間または属性検索を行うことで総合的分析を進めていくケースがほとんどである。このデータの重ね合わせの状態をレイヤと呼ぶが、GISの操作では、このレイヤをいかに効率よくビジュアル化するか、ということが重要なポイントとなっている（Fig.8）。

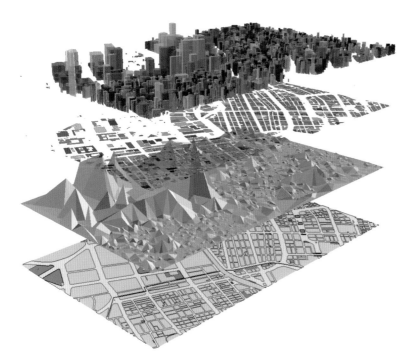

（Fig.8）レイヤの概念図
さまざまな地域分析図を層状に重ねて表現することで、今まで見えなかった地域の
特徴や課題が明確に把握される

CHAPTER 1, 4 VISUALIZING THE CITY USING GEOGRAPHIC INFORMATION SYSTEM

Visualization of Urban Layers

■ What is the role of an effective geographic information system (GIS) in city planning?

■ Linking map information with Excel information for spatial information analysis

■ Visualize a city by converting 2D maps to 3D according to the purpose

A GIS that is capable of digitizing maps and processing large volumes of data

The Geospatial Information Authority of the Ministry of Land, Infrastructure, Transport, and Tourism defines GIS as "a technology that enables advanced analysis and rapid decision-making by comprehensively managing, processing and visually displaying data with spatial data based on geographic locations." There are two types of spatial data: data recorded in the symbols themselves plotted on a map and information recorded in a square grid of the same size on which the map is divided. The former is treated in GIS as vector data and the latter as raster data, which are different data systems. Generally, the term "GIS" is often used to refer to the map itself, which contains information; however, in this section, it is defined in a broad sense that includes maps.

GIS can be used as an information analysis tool

Currently, GIS is being used extensively in general life due to the spread of smartphones and other factors.

(1) Use as a map: GPS-linked navigation systems, etc.
(2) Use as an information stocking tool: digitized paper-based urban planning maps, etc.
(3) Use as an information browsing tool: map browsing through search engines, etc.
(4) Use as a communication tool: Interactive exchanges such as "Pokémon GO"
(5) Utilization as an information analysis tool: effective application to urban design and city planning, etc.

In particular, 5) is extremely important for those of us involved in architecture and urban planning when conducting spatial analysis. First, GIS's compatibility with Excel data makes it possible to perform statistical analysis and, by linking with statistical analysis software such as SPSS, it is possible to perform large-volume multivariate analyses. Another analytical technique unique to GIS is its ability to perform spatial analysis by processing not only tabular data but also map data. This is an advantage of GIS, which can stock location information as visual information. Below we introduce some of the specific applications of GIS data.

Examples of GIS Data Applications

Data handled in GIS can be divided into three main types: (1) vector data, which is data about the elements plotted on a map (hereinafter referred to as "features"), (2) raster data, and (3) image data. Generally, data such as photographic images are composed of pixels and are often treated as raster data. For the sake of convenience, image data will be unified here as image data, and map information on a grid with Excel table data will be called raster data.

As the name suggests, the components of GIS are maps with Excel table data that complement the map information. Map information is used to express specific locations and, in the case of vector data, sizes and shapes, and the information attributed to each map element is recorded as Excel table data, which correspond to the features on a one-to-one basis where each element is automatically assigned an ID and represented as a row in the Excel table data (Fig.7).

The relationship between map information and Excel table data means that all information can be statistically analyzed using Excel. It is also easy to create a specific color-coded representation of the data created by the new statistical analysis. Thus, when map information and Excel table data are related, spatial evaluation through statistical analysis can be easily performed, and GIS is an extremely effective tool for strengthening the communication of survey results and the persuasive power of design concepts. In actual GIS analysis work, a comprehensive analysis is conducted by overlaying several files with several conditions and performing spatial or attribute searches. This layering of data is called a layer, and an important point in GIS operation is how to visualize the layers efficiently (Fig.9).

Fig.1   Example of a plot of open space in Tokyo
Fig.2   Representation of road types by thickness and color based on the purpose
Fig.3   3D representation of a city using polygon data
Fig.4   Elevation representation by continuous raster data (left) and land use map by discrete raster data (right)
Fig.5   Color-coding of elevation converted from continuous raster data to discrete raster data
         Map information and Excel table data
Fig.6   One-to-one corresponding features and Excel table data (Vector data)
Fig.7   Features and Excel table data (raster data) corresponding one-to-one
Fig.8   Image of Urban layers

Chapter 1 / 5 / 市民参加による「まちづくりデザイン」
1. アーバンインターベンションとシャレットワークショップ

- ■ 調査内容を「見える化」して
  新しいコミュニケーションを生む
- ■ 集中デザインワークショップで
  市民の想いを形にする
- ■ 将来のビジョンを共有し、地域の魅力を
  最大限に引き出す

アーバンインターベンションの考え方

地方都市における「まちづくりデザイン」で重要なのは、科学的あるいは中立的な情報を行政や専門家が市民に提供し、共有された情報の中から重要な課題を抽出して、将来的にその課題をどのように解決し、どう地域に導いていくべきかという「将来ビジョン」を共有することである。ビジョンが共有されれば、それを具体的に専門家や行政の人たちが加わって、短期、中期、長期的計画の中に落とし込み、予算をつけて実現するという具体的なプロセスにつながりやすくなる。このときに、都市の一部分である建物や広場などが、地域全体にどのような影響を及ぼすかを考えることは計画する者にとっては重要な視点となる。具体的には、将来ビジョンを実現するために、都市の部分的要素が全体に与えるインパクトを予想し、その方法を具体的に考察することが求められている。この考え方は、医学の領域でも使われている「インターベンション（介入）」の考え方に近い。

医学では、人間の体を診断し、疾患内容を把握して治療方法を考え、具体的な治療および投薬を行うが、都市においては、「アーバンインターベンション（都市介入）」という言い方をし、調査による診断を通して課題を把握し、「ツボ」を押して血流を良くするように、建物や広場を再編成することで、地域の活性化を目指すことが海外では頻繁に行われている。都市計画が俯瞰的に都市をとらえて、制度や土地利用を作成するのに比べて、「まちづくりデザイン」や「アーバンインターベンション」は空間をデザインする建築系人材こそができる重要な仕事である。しかし、都市的スケールの問題について、異なった専門領域を持った個人のデザイナーや技術者ができることには限界があるため、現実的には以下で示す「シャレットワークショップ」のようなプロセスで、専門家集団によるまちづくりデザインの支援が行われることが多い。

シャレットワークショップによる実践

ヨーロッパや米国のコミュニティ計画における主要な手法として取り入れられているワークショップの手法に、専門家集団による集中的ワークショップを意味する「シャレットワークショップ」がある。通常、1週間程度の短期間に、20人近くのさまざまな領域の専門家が現地入りし、行政や住民と不定期に会合を重ねながら、具体的な計画案を示し、何回も議論を繰り返しながら最終的な合意案（マスタープラン、主要なランドスケープ、デザインコード、典型的な建物の計画など）を確定する。そもそも「シャレット」という言葉の原義は、フランス語で荷馬車という意味であったが、パリ美術学校（エコール・デ・ボザール）の作品提出締め切り直前の駆け込みを起源に「短期間に駆け込みで計画を行う」という意味に変化したと言われる。

一般にわが国で普及しているワークショップの形態は、主として地理学者、文化人類学者の川喜多次郎（1920〜2009年）により開発されたKJ法や、オーストリア出身の都市計画家・建築家クリストファー・アレグザンダー（1936〜2022年）によるパターンランゲージという手法を用い、住民に自ら手を動かしてもらい、問題点をあぶり出し、意識を共有するものが多い。「シャレットワークショップ」は、これらと混同されやすいが、専門家集団により、具体的な解決策を市民に提示し、なぜそのような解決策になったかというプロセスを説明するところが大きな違いである。特に公共事業の場合には、市民の合意形成は、その場に参加した関係者の共有感だけでは不足しており、多くの人々への説明責任を果たせるものでなくてはならないからである。

ワークショップの具体的な進行については以下の段階を踏む（Fig.1）。

基本情報の収集	対象敷地の視察	市民の意見収集	敷地と周辺の分析	問題点の把握	将来ビジョンの検討	具体的なデザイン	市民・行政への発表

（Fig.1）シャレットワークショップのプロセス

主要なプロセス

1）基本情報の収集（目的とプロセスの把握）

シャレットワークショップの位置づけと目標を明確にするための基本情報を収集する。具体的には、①地域がもつ独自の文化的地理的特徴を文献や実際の街から読み取り、いかにそれを強化するかを考え、②地域の課題を解決するための方策を考えるとともに、現実的な制約条件の中でいかに実現できるかを予め考える。

2）対象敷地の視察（まち歩きによる地域の読取り）

まずは対象地の概要を把握するために、実際にまちを歩き回って、肌感覚で現地調査をする。調査方法には多様なレベルがあり、①あまり時間をかけないで直観的に対象地区の特徴（強み、弱み）を読み取り把握する作業と、②詳細な調査情報をもとに、じっくり時間をかけて科学的に対象地区を調べ記録する作業を、明確に分離して位置づけることが重要である。

3）地元の意見収集（市民へのヒヤリング）

行政による既存の都市計画方針や上位の政策は、文献やガイダンスを通じて把握することは可能であるが、ステイクホルダーと呼ばれる市民たちから現状に対する意見やあるべき姿のイメージなどを引き出し、計画策定のための根拠を考えることはワークショップの成否を決定するほど重要なポイントである。最初のヒヤリングと中間発表時における市民の意見のフィードバックは、ワークショップの提案を現実的なものとして位置付けるための必須の作業項目となる。

4）敷地と周辺の分析と将来ビジョンの策定

対象地区とその周辺地区の形状、歴史、街並みなどを改めて調査分析し、その地区が抱える「問題点」を確認するとともに、「可能性」のある方向性を模索する。このプロセスは地区の特徴を深く読み込み、それをどのように強化すべきかであるかを検討する基礎となる。最終的に、その可能性を何に活かすのかという「将来ビジョン」を文章やイメージの形で想定する。

5）具体的な空間デザイン

想定された「将来ビジョン」をもとに、最も効果的だと思われる具体的な場所を採り上げ、コンテクストに配慮した空間デザインを提案する。公共空間の価値を上げるための作業提案と言い換えることもできる。

具体的には、1）地域課題の提示とその改善方法の提案、2）模型および3D画像による改善後のイメージの提示、3）景観シミュレーションなどによる将来景観イメージの提示、4）これらの計画による具体的な居住環境の改善や経済効果の提示、などである。この最終プロセスは、アーバンインターベンションによる具体的な処方の提案であり、針灸治療におけるツボに類似し、重要な場所に操作を行うことで、地区全体に良い影響が生まれることを見込む作業である。

6）市民への発表と記録および広報

最終的に、検討の成果を市民や行政関係者に対してプレゼンテーションする。具体的なデザインを一過性のものとして終わらせないために、即座にデジタル化して記録として残すと共ともに、出来るだけ多くの市民に周知するために、地元のケーブルテレビや地域新聞などのメディアを通して広報することも重要である。

以上が、シャレットワークショップのプロセスの概要である。これらのプロセスについて、各段階における教育効果をまとめたものが以下のダイアグラムである（Fig.2）。

作業のプロセス	作業内容	教育的効果
予備調査	大まかな問題点の抽出 地方都市共通の問題の洗い出し	歴史的資産の保存に関する活動や制度の確認 「まちづくり」全般への理解
基本情報の入手と診断	行政・住民からの事情聴取 現地の景観調査、建物の実寸調査など	フィールドサーヴェイを通じ、抽象的論議を 具体的な課題として自分の眼で確認する
各種のシミュレーションと 解決策の模索	CGによる景観シミュレーション／ 模型やスケッチによる具体的将来ビジョンの提案／ 解決のためのシナリオ作成／代替案による問題の顕在化／ ブレーンストーミングによる解決案の模索など	グループ別の作業やディスカッションを通じ、最良の解を 求めると同時に、多角的な解の存在を認識する／ 短時間におけるビジョン確定と模型の制作 （日頃の技術の応用）
関係者への プレゼンテーション	診断結果から導いた解法ビジョンの幾つかについて行政や 住民に公開し、意見を求めると共に関心を高めてもらう	クライアントやユーザー、コミュニティーのメンバー等を 想定した複数の相手に対するプレゼンテーションの訓練
記録・広報	ケーブルテレビ、広報誌などのメディアを用いた 活動内容の外部発信	成果の整理、活動内容についての反省などのフィードバック／ 対外的な評価の確認／今後の活動方向の確認など

（Fig.2）「シャレット・ワークショップ」の段階的フローとその教育的効果

"Urban Intervention" and "Charrette Workshop"

■ Visualize the survey reports to generate new communication

■ Intensive design workshops to shape citizens' thoughts and ideas

■ Share a vision for the future and maximize the potential of the community

What is "Urban Intervention"?

In "community development design" in local cities, it is important for the government and experts to provide citizens with scientific or neutral information, identify important issues from the shared information, and share a "vision for the future" on how these issues should be solved and how the region should be managed in the future. Once this vision is shared, it will be easier to connect it to the concrete process of incorporating it into short, medium, and long-term plans with the participation of experts and government officials, and then budgeting and implementing those plans. At this point, planners need to consider how the existing buildings and plazas in the city affect the entire community. Specifically, the planners are required to consider the specific ways in which the components of the city can anticipate their impact on the whole city in order to realize their vision for the future. However, there are limits to what individual designers and planners can do; therefore, in reality, community development design is often supported by a process of "charrette workshops" provided by a group of professionals.

Charrette Workshop in Practice

The "charrette workshop," which means an intensive workshop by a group of experts, is one of the main methods used in community planning in Europe and the United States. In a short period, usually about one week, nearly 20 experts from various fields visit the site, meet irregularly with the government and residents, present concrete planning proposals, and finalize a consensus plan (master plan, major landscapes, design codes, typical building plans, etc.) through repeated discussions. The process aims to finalize the plan. The workshop follows the following steps:

Main process

(1) Gathering basic information: understanding the purpose and process

Basic information is collected to clarify the positioning and goals of the charrette workshop: a) Read literature on the unique cultural and geographical characteristics of the region and the town and suggest ways to enhance them. b) Suggest measures to solve local problems and how they can be realized within realistic constraints beforehand.

(2) Fieldwork: The community walks around the town

First, it is important to actually walk around the city to get an overview of the target site and conduct an on-site survey. Survey methods have various levels, and it is important to separate and position (a) the process of intuitively reading and understanding the characteristics (strengths and weaknesses) of the target area without spending too much time on it, and (b) the process of taking the time to scientifically investigate and record the target area based on detailed survey information.

(3) Understanding local opinions: interviews with citizens

It is important to obtain citizens' or stakeholders' opinions on the current situation and the future image of the area. These opinions become the basis for the planning process, which determines the success or failure of the workshop. Feedback from the public during the initial interviews and interim presentations is an essential part of the process to ensure that the workshop proposals are realistic.

(4) Analysis of the site and surrounding area, identification of problems, and consideration of future vision

The shape, history, streetscape, etc. of the site and its surrounding areas are again surveyed and analyzed to identify the problems and seek possible directions. Finally, a "vision for the future" regarding what to do with the possibilities is envisioned in the form of text and images.

(5) Space design

Based on the "vision for the future," specific locations that are thought to be most effective are selected, and a detailed spatial design is proposed to increase the value of public space.

The process includes: a) presentation of the local issues and their improvement methods, b) presentation of the image after improvement using physical models and 3D images, c) presentation of the future landscape image using landscape simulation, etc., and d) presentation of the specific improvement of the living environment and economic effects of these plans.

(6) Presentation to the public, documentation, and publicity

Finally, the results of the study are presented to citizens and government officials. To ensure that the specific design does not end up as a one-time event, it is important to immediately record and digitize the results and publicize them through media such as local cable TV and regional newspapers to inform as many citizens as possible.

The above is an overview of the charrette workshop process. The following diagram summarizes the educational effects of each stage of the process.

Fig.1   Charrette workshop process
Fig.2   Charrette workshop process and its educational effects

市民参加による「まちづくりデザイン」
2. 将来ビジョンの共有とエリアマネジメント

- ■ 「まちづくり」から「まちの運営」へシフトする
- ■ 企業型エリアマネジメントから地域主導の
  エリアマネジメントへ
- ■ 「つくる」から「つかう」への意識の転換が求められる

わが国が、高度経済成長から人口減少時代へと社会情勢が変化するにつれ、まちづくりデザインの様相や役割は変わってきた。都市そのものを新たにデザインするというよりは、① 都市のあるべき将来像をビジョンとして市民と共有する、②まちづくりを実現するためのアクションプランを市民参画で考え、具体的に街路や広場などの公共空間を魅力的に再編成し、都市生活やライフスタイルを豊かにする、③それらのデザインプロセスを市民へ公開しながら合意形成を継続していく、という3つの役割が強くなってきた。また同時に、地域を長期的に運営する仕組みが必要になっており、そこに最もエリアマネジメントが注目されるポイントがある。

エリアマネジメントとは何か?

エリアマネジメントの定義は、「地域の価値を維持・向上させ、また新たな地域価値を創造するための、市民・事業者・地権者などによる絆をもとに行う主体的な取り組みとその組織、官民連携の仕組みづくりのこと」(法政大学 現代福祉学部・人間社会研究科 教授 博士(工学) 保井美樹)であるといわれる。しかし、現在、全国各地でエリアマネジメントに向けた活動が展開しているが、「エリアマネジメント」という言葉の意味は、各地で異なっている。原点に立ち返ってみると、重要なのは、(1)地域住民や行政が、自分たちが住み、働く町の特徴や魅力を把握し、共通の将来ビジョンをもち得ているかという点であり(2)運動論として、その共有されたビジョンを実現するための方法論が考えられているか(3)そして組織論としてどのような組織体を構築し、継続的に運営していくか、ということである。

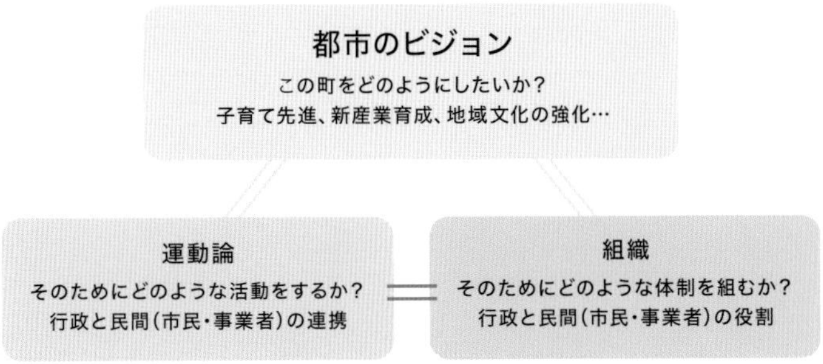

都市のビジョン
この町をどのようにしたいか?
子育て先進、新産業育成、地域文化の強化…

運動論
そのためにどのような活動をするか?
行政と民間(市民・事業者)の連携

組織
そのためにどのような体制を組むか?
行政と民間(市民・事業者)の役割

(Fig.1) エリアマネジメントの抱えている問題点
エリマネ組織のあり方や財政面の問題など教科書がない状態が続いている

特に(1)は、多くの場合、共有のプロセスが不十分な場合が多い。(Fig.1)はその関係を示したダイアグラムである。

これまでのエリアマネジメントと
新しいエリアマネジメントのフェーズ

エリアマネジメントという言葉の概念は、ここ20年程度でようやく浸透してきたが、わが国では、地域の共同体で地域をマネジメントすることは、江戸時代からの火消に代表されるように、すでに300年前から小規模の自治組織に存在していた。現代では、町内会や商店街組織がそれに代わった地縁組織として存在し、アーケード整備や管理、集客イベントなどを会費や補助金で実施してきたのである。今までは行政が都市を統制し、「管理」することが当たり前になっていたが、いよいよ経済衰退や人口減少に伴い税収が落込むなかで、行財政の逼迫から都市地域を行政だけで維持することが困難になってきた。それに対し、地域の地権者や市民が自ら資金を出し、「まちづくり会社」などを設立することで、まちづくり事業により収益を得て、地域に必要な事業を実施することで地域の価値を高めていく、いわゆる「エリアマネジメント」が必要になってきたのである。

わが国のエリアマネジメントの第1フェーズは、2002年に設立された「NPO法人 大丸有エリアマネジメン協会」に端を発した「再開発型エリアマネジメント」である。その後、都内では汐留シオサイトや大崎西口地区、東五反田地区など、再開発地権者からの負担金や人的持出しに依存した仕組みで、主にエリアプロモーションや公共施設の維持管理を主な事業とする形で行われてきた。毎年企業側のメンバーに異動があるなど、人間関係が安定している住居系の地域におけるエリアマネージジメントとは様相が異なっている。その後、第2フェーズに入り、2009年頃から地方中枢都市を中心として財源確保策を工夫し、駅を中心とした中心市街地再生を対象に「地方都市型エリアマネジメント」の事例が模索され始めた。このマネジメントスキームの成立要件には、公共空間(パブリックスペース)の活用のあり方が変化したことや、その重要性を理解した市民のパブリックライフへの関心が強くなったことが背景に存在している。

公共空間の戦略的活用を基軸とした
エリアマネジメントへ

2011年の「都市再生特別措置法」の改正により、「道路占用許可の特例」制度が創設された。これは、道路空間に一定の歩行者通行機能と安全性を確保したうえで、常設の飲食・物販店舗やオープンカフェ、サイクルポートなどを「道路占用許可の特例」を基に設置し、公共空間から得られる収益をまちづくり財源や道路の維持管理費等に充てることができるという仕組みである。東京の「新宿モア4番街」を全国初の事例として、札幌大通地区、大阪グランフロント、鳥取・バード・ハット、高崎

市、岡山市、柏の葉キャンパス駅西口、虎ノ門ヒルズ・新虎通りなどがそれに続いた。このように制度改正や社会情勢の変化、市民の成熟などから、駅前広場や鉄道跡地の空地活用などを含めて、全国で公共空間の使い方やあり方の見直しが始まったのである。特に地方都市のエリアマネジメントにおいては、それが重要な流れとなっており、エリアマネジメントの対象エリアが、民有地とともに公共空間（街路、駅前広場、河川、公園等）が多くを占めるため、その公共空間をいかに魅力的にして市民が集う場にするか、そしてその場でのコミュニティ形成やソーシャルキャピタル（社会関係資本）の形成にいかに寄与し、中心市街地再生に間接的にも相乗効果を与えられるかどうかは重要なポイントである（Fig.1）。

「つくる」から「つかう」へ

現代の「エリアマネジメント」を概観すると、当初の第1フェーズから、第2フェーズを経て、すでに第3のフェーズに入ってきているといえよう。
従来の「まちづくり」は道路などのインフラ整備に加え、商業施設オフィス、マンションなどの新しい開発の側面が強い「つくる」まちづくりであった。それは、行政、ディベロッパー、不動産業界を中心に、ハード

的側面により加速度的に都市の成長が促進された。一方、自治体の画一的な行政サービスは、財政難の中で少子高齢化が進む社会に対し、きめ細かい市民サービスなどに対応できないため、まちの活性などのソフト面に寄与する新しい「仕組み」が求められるようになった。このような背景をもとに、そのエリアが抱える問題の解決と、解決後の良質な生活の質（QOL）を維持し、今ある資源を「つかう」ためのエリアマネジメントに注目が集まってきた

国土交通省はウォーカブル（歩きやすい）な都市を推奨し、各自治体では、「つくる」から「つかう」への意識転換が少しずつ進んでいる。特に、あまり使われていない公共空間を民間の力で生き返らせ、歩いて15分や20分の距離内で生活に必要なさまざまが足りる職住近接型のまちづくりの考え方が、アメリカのポートランド、フランスのパリ、オーストラリアのメルボルンなどで強力に推し進められている。民間の力で、まちを「つかい」ながら運営し、歩いて安心に暮らせるコンパクトな界隈をつくることが世界中で求められているのである。そのための、ICT（Information and Communication Technology）も含めた新しいエリアマネジメントのシステムづくりが今後の日本にもおいても重要な課題となっている。

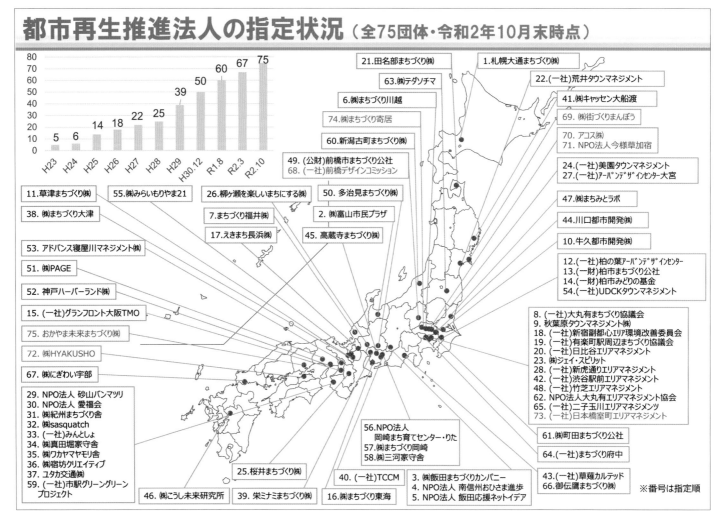

（Fig.2）全国における「都市再生整備法人」の普及状況（国土交通省HPより）
ほとんどの都道府県に、さまざまな組織形態の法人が生まれており、地方では公共空間における「まちおこし」が期待される

Sharing a Vision for the Future and Managing the City

- ■ Shift from community development to city management
- ■ From corporate-oriented area management to community-led area management
- ■ A shift in mindset from "create" to "use"

As the social situation in Japan has changed from an era of rapid economic growth to a declining population, the aspect and role of urban development design have changed. Rather than designing the city itself anew, the role of urban development design has changed to: 1) sharing the future vision of the city with citizens; 2) involving citizens in action plans to realize urban development, specifically reorganizing public spaces such as streets and squares into more attractive places, and enriching urban life and lifestyles; and 3) continuing to build consensus while disclosing the design process to the public. At the same time, a mechanism for the long-term management of the region is needed, and this is the point where area management is attracting the most attention.

What is Area Management?

Miki Yasui (Hosei University) defines area management as "a proactive approach and organization based on ties among citizens, businesses, landowners, and others to maintain and improve the value of an area and to create new regional value, as well as a mechanism for public-private partnerships." However, while activities for area management are currently being developed in various parts of Japan, the meaning of the term "area management" differs in various regions. Going back to the starting point, what is important is (1) whether local residents and administrators have a common vision for the future based on an understanding of the characteristics and attractiveness of the towns or cities where they live and work, (2) whether a methodology for realizing that shared vision is being considered as part of active movement, and (3) what kind of organizational structure is needed to realize that shared vision and how will it be managed on an ongoing basis?

Regarding (1), this sharing process is inadequate in many cases. Fig.1 shows this relationship.

Previous area management and new area management phases

Although the concept of area management has only become widespread in the last 20 years, in Japan, local community-based area management has existed for 300 years in small-scale self-governing organizations, as exemplified by fire extinguishers from the Edo period (1603-1868). In modern times, chonai-kai (neighborhood associations) and shopping district organizations have existed as local community organizations, and arcade maintenance, management, and events to attract visitors have been carried out with membership fees and subsidies. Until now, it has been the norm for the government to control and "manage" the city; however, with tax revenues declining due to economic weakening and a shrinking population, it has become difficult for the government alone to maintain the city under tight administrative and fiscal conditions. In response, it has become necessary for local landowners and citizens to finance and establish "community development companies" to generate revenue through community development projects and to enhance their value by implementing necessary local projects.

The first phase of area management in Japan was "redevelopment-type area management," which originated with the "NPO Daimaru Area Management Association" established in 2002. Subsequently, in Tokyo, the Shiodome Sio-site, Osaki West Exit Area, and Higashi-Gotanda Area have been conducted under a system that relies on contributions and human resources brought in by redevelopment landowners, mainly in the form of area promotion and the maintenance and management of public facilities. The system differs from area management in residential areas, where human relations are more stable, as members of the corporate side are transferred every year. In the second phase, around 2009, local cities began to devise measures to secure financial resources, mainly in regional cities, and began to seek examples of "local city-type area management" targeting the revitalization of central city areas centering on train stations. The requirements for the establishment of this management scheme exist in the background of changes in the way public space is utilized and a stronger interest in public life among citizens who understand its importance. In an overview of contemporary "area management," it is fair to say that from the initial first phase, we have already passed through the second phase and are now entering the third phase.

Toward Area Management Based on Strategic Use of Public Space

Under these circumstances, the 2011 amendment to the Law on Special Measures for Urban Revitalization established the "Special Exception for Road Occupancy Permission" system. Under this system, permanent food and beverage outlets, open cafes, cycle ports, and other facilities can be set up on the basis of "special exceptions for road occupancy permits," and the revenue generated from the public space can be used to finance urban development and road maintenance costs. The "Shinjuku Moa 4th Avenue" in Tokyo was the first case implemented, followed by many other cities across Japan. Thus, following institutional reforms, changes in social conditions, and the maturation of citizens, a review of public space usage began throughout Japan, including station plazas and vacant land on former railroad sites. This is a particularly important trend in the area management of local cities, where public space (streets, station squares, rivers, parks, etc.), along with privately owned land, make up a large portion of the target area of area management. Therefore, it is important to consider how to make such public spaces attractive and a place where citizens can gather, as well as how to promote community bonding and social capital in such places.

From "Create" to "Use"

Conventional "urban development" has been a "building" process, with a strong emphasis on the development of new commercial facilities, offices, condominiums, etc., in addition to the construction of roads and other infrastructure. This has accelerated the growth of cities through tangible aspects, mainly by governments, developers, and the real estate industry. However, the uniform administrative services of local governments have been unable to provide detailed services to citizens in a financially strapped society with a declining birthrate and an aging population. Therefore, new "mechanisms" that would contribute to the intangible aspects of urban revitalization are required. Area management has been attracting attention as a means of solving the problems faced by the area and maintaining a good quality of life "using" the existing resources.

The Ministry of Land, Infrastructure, Transport and Tourism has recommended walkable cities, and local governments are gradually shifting their focus from "create" to "use." In particular, the concept of compact city planning, in which underutilized public spaces are revitalized by the private sector and various live-work-play neighborhoods can be completed within a 15- or 20-minute walk. The creation of a new area management system, including the integration of ICT, will be important for Japan in the future.

Fig.1   Important elements that make up area management

Fig.2   The prevalence of "Urban Revitalization Corporations" throughout Japan (from the website of the Ministry of Land, Infrastructure, Transport and Tourism)

URBAN VISUALIZATION

CHANGES OUR TOWN

SIMULATION

CHAPTER 2

第二章　10のシミュレーションで都市を「見える化」する

街並みの形態における交換テスト
景観の価値を見える化する

概要

　一章（10頁）で示した記号論の考え方を応用し、都市や地域の景観のアイデンティティーについて考察し、ツールの可能性を探る。日本の街並みは、元来その地域独自の特色を反映して育成されてきた。その特色には歴史、文化、産業、風土、素材などの地域特性がある。それぞれの都市や地域の景観における固有のアイデンティティーは、その地域特性を反映したさまざまな景観要素から成り立っており、それらの景観要素の何がその地域のアイデンティティーを最も強く表現しているのか、すなわち、景観のアイデンティファイアー（identifier）と呼ばれる要素を抽出するために、言語学で利用している交換テストを具体的な景観の分析方法として応用する。

景観の価値評価方法について

　景観の価値評価とは、それが持つ固有のイメージしやすさ（image-ability）や魅力度（attractiveness）を客観評価することであるが、これについては今まで主観的評価対象として扱われ、具体的な科学的論議になりにくい課題であった。ここでは、共通の価値観として、景観に魅力度が存在するという前提で議論を進める。景観は、それを構成している諸要素（＝景観構成要素）の影響によりイメージしやすさが変化し、構成要素の相互関係により景観の総合的な魅力度の増加・減少という変化が生じる（Fig.1）。また構成要素ごとに魅力度の要因として占めている割合が異なると考えられ、全体の魅力度に対して高い影響力を与えている構成要素を基準にすると、魅力度へ貢献する要素間の差を設定できる。最終的には、全体の中での景観構成要素間のヒエラルキーを明確にすることで、個々の景観の価値評価が可能になることを確認する。
　そこで、ここでは景観の価値評価を実践していくうえでの第一歩として、下記の2点を試みる。
1）景観の魅力度に対して高い影響力をもつ主要構成要素の抽出
2）構成要素間の魅力度に対する要因の差＝「重み」の算出。

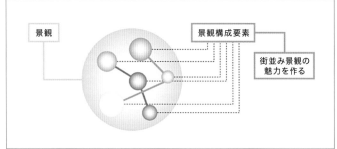

（Fig.1）景観の構造
景観構成要素の状態により、魅力の増加・減少という変化が生じる

　2）の重みの抽出方法については、景観Sにおける構成要素：Xnの景観の魅力度に対する影響度を抽出する例でこれを示す（Fig.2）。景観Sから構成要素Xnを消去して【景観S-Xn】を作成し、【景観S】と【景観S-Xn】の景観的魅力の差を抽出する。この魅力度の変化量は構成要素：Xnが引き起こす魅力度の変化量であり、構成要素Xnがもつ景観的魅力に対する影響度であると考えられる。以上の方法により各景観構成要素の、景観の魅力度に対する影響度を算出することが可能となる（Fig.3）。

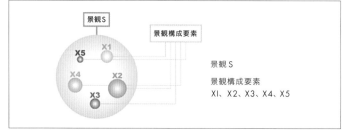

（Fig.2）景観構成要素と景観Sとの関係

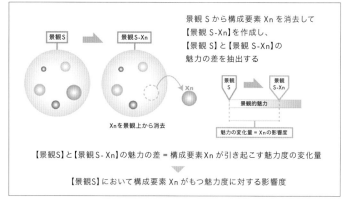

（Fig.3）交換テストによる景観構成要素の評定

SD法と重回帰分析による主要景観構成要素の抽出

　分析方法については、印象調査を行う必要があるため、SD（Semantic Differential）法と重回帰分析の手法を使う。まず、以下のプロセスを行う。
1）景観写真を用いた意識調査と数値解析
2）3D景観画像によるシミュレーションによる印象調査
　分析対象地は、既存の価値観の影響が顕著に現れていると考えられる商業地を選定し、都内4ヵ所（銀座、表参道、下北沢、代官山）の商業ストリートで分析を行った。
　1）では、景観写真を用いたアンケート調査を、被験者31名を対象に実施した。まず景観の魅力値の把握では個々の対象景観がもつ景観の魅力を、SD法により抽出し数値化した。ここでは5つの質問事項

に対して6段階の回答を得て、5項目の平均値を魅力値と設定している。

（Photo 1）対象地の景観

次に主要構成要素の抽出では景観写真より各景観（A～D）の良い・悪いに関係していると感じる構成要素を指摘してもらう。基本的な指摘方法としては事前に列挙した構成要素から選択してもらい、それ以外のものについては記述を依頼している（Fig.4）。

【SD法質問事項】

おもしろい ― おもしろくない
魅力のある ― 魅力のない
好ましい ― 好ましくない
良い ― 悪い
楽しい ― つまらない

【各景観の魅力値】

A：銀座	B：表参道	C：下北沢	D：代官山
3.460	4.053	4.300	4.067

（Fig.4）SD法による評価値

 X1 建物高さ  X2 色彩

 X3 開口部 X4 看板・サイン

 X5 電柱・電線 X6 道路舗装

 X7 樹木 X8 街灯

（Photo 2）主要構成要素の現れ方

建物1次要素： 建物高さ 色彩 材質 屋根
　　　　　　 庇 窓・開口部 柱 壁
　　　　　　 格子 ドア・戸
建物2次要素： 門 柵 塀 看板・サイン
街路要素： 電柱・電線 道路舗装
　　　　　 樹木・植栽 街灯 道路標識
　　　　　 道路標示 自動販売機

（Fig.5）主要構成要素の抽出

主要景観構成要素

建物要素群	街路要素群
X1：建物高さ	X5：電柱・電線
X2：色彩	X6：道路舗装
X3：開口部	X7：樹木・植栽
X4：看板・サイン	X8：街灯

（Fig.6）抽出された主要構成要素

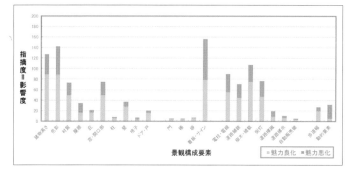

（Fig.7）全景観構成要素の景観魅力に対する影響度グラフ

アンケートで抽出できた、景観の良い・悪いに対する各構成要素の指摘数の総和を、現状の景観の魅力度に対して構成要素が保持している影響力ととらえ、4つの景観（A～D）において、魅力度に対して総合的・平均的に高い影響度を示している構成要素を、主要景観構成要素に選定した結果、これらの要素が抽出された（Photo2）（Fig.5）。

要素の重み付けでは先の意識調査より抽出したデータをもとに
- 景観の魅力値＝目的変数
- 主要景観構成要素の魅力に対する影響度＝説明変数
に設定し、重回帰分析により構成要素間の重み付けを行う。
（説明変数を【建物要素群】と【街路要素群】に分けて分析する。）
重回帰分析により求められた、
- 解析の精度を表す『重相関係数R』
- 説明のウェイトを表す各構成要素の『重回帰係数』
より構成要素間の重みを算出する。
【重み】＝｜重回帰係数｜× 重相関係数R × 補正係数（10）

X1	X2	X3	X4	X5	X6	X7	X8
0.573	0.528	0.176	0.321	0.921	0.688	0.704	0.440

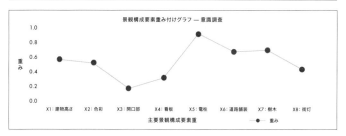

（Fig.8）主要景観構成要素重み付け ― 重回帰分析

これらのデータをもとに、再度構成要素の有無によるシミュレーション画像を作成し、被験者の印象調査を繰り返す作業を行った。以下は、銀座、下北沢における交換テストの事例である（Fig.8, 9, 10）。交換テストを用いた街並みの再構成については、実際に景観の修景を検討する際に、修景の方向を探るための合意形成プロセスの中で有効なツールになると思われる。

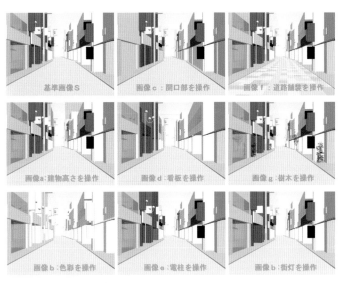

（Fig.9）下北沢における交換テスト
道路の舗装、電柱、看板、色彩などにより、通りの雰囲気が変わるのが分かる
下北沢らしさを醸し出している景観構成要素は明らかに銀座や代官山とは異なっている
そこに景観のアイデンティティーのキーがある

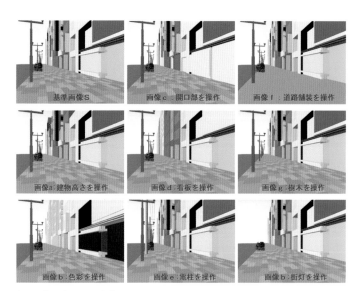

（Fig.10）銀座における交換テスト
看板、街頭、色彩などにより、景観の雰囲気が変わるのがよくわかる

TEST 1: 現状

TEST 5: 屋外広告物規制

TEST 2: 高さ規制

TEST 6: 電柱・電線規制

TEST 3: 意匠操作による修景

TEST 7: 道路舗装規制

TEST 4: 色彩規制

（Fig.11）交換テストにおける景観構成要素のマトリックス

日本橋人形町における景観の交換テストの事例

　ここでは、2004年に施行された「景観法」による地方自治体の景観
規制の考え方を念頭に、様々な景観要素の有無をデザインコード化
（景観形成のためのルール）することを検討した事例を示す。具体的に
は、歴史的雰囲気を残しているが、あまり景観政策に対し積極的では
ない地域に対し、街並みの交換テストによる景観シミュレーションを
行い、行政や市民の意識を促し、景観デザインガイドライン（建物の外
観デザインを規制するためのルール）を策定するための根拠をつくる
検討を行った。対象地としては、「日本橋人形町」をモデルに「人形町ら
しい」と感じるためにどの要素が重要な役割を果たしているかを検討
するための画像を作成している。実際に立面のデザインガイドライン

の規制をかけることを想定し、何が「日本橋人形町らしさ」であるのか
を探ったが、結果としては、建物の高さ、色彩、電柱、看板などが景観
の魅力を左右する強い要素としてあげられた（Fig.11）。

操作実行要素

要素＼タイプ	TEST 1	TEST 2	TEST 3	TEST 4	TEST 5	TEST 6	TEST 7
現状	●	●	●	●	●	●	●
建築物等に関する規制							
建物高さ		●					
形態		●	●				
意匠			●				
色彩				●			
工作物に関する規制							
屋外工作物					●		
電柱・電線						●	
道路							●

（Fig.12）日本橋人形町における交換テスト

　多様な景観要素が複雑に絡み合って表現されている「地域らしさ」
の構造について、景観の構成要素の面から少し踏み込んで分析した。
あるレベルまでは、景観の再現性の可能性に言及できるが、実際の地
域景観には、「生活景」と言われるように、地域における「さまざまな生
業」、「生活感」、「香り」、「音」など視覚以外の環境要素があるので、視
覚だけで完全に再現性にたどり着くわけではない。ここでの分析成果
にさらに他の分析方法を重ねることが重要である。

Visualization of the Townscape Value

SUMMARY

Applying the ideas of semiotics presented in Chapter 1 (p.10), this chapter discusses the identity of urban and regional townscapes and explores the possiblepossibilities of investigation tools. Japanese townscapes are traditionally developed to reflect the unique characteristics of the their region, that including itses history, culture, industry, climate, materials, and other features, which can be understood as elements of regional characteristics. Therefore, in order to determine and extract the so-called "identifier" of a townscape (elements that most strongly expresses the identity of the region), we apply the exchange test used in linguistics as a method of analyzing specific townscapes.

Evaluating townscape value using the exchange test

The evaluation of townscape value involves the objective evaluation of townscapes' inherent imageability and attractiveness. However, this has typically been treated as a matter of subjective evaluation and is difficult to discuss in a concrete scientific manner. Here, we will proceed with the assumption that attractiveness exists in townscapes as a common value. The imageability of a townscape changes according to the influence of various elements that make up the townscape (i.e., townscape components), and the interrelationship of the components causes changes in the overall attractiveness. The proportion of each component as a factor of attractiveness is considered to be different, and we will attempt to establish the differences between those that have a high influence on the overall attractiveness. Finally, we confirm the value evaluation of individual townscapes by clarifying the hierarchy of townscape components amongst the whole.

Therefore, as a first step in practicing townscape value assessment, we attempt here the following two points.
> (1) Extraction of the main components that have a high influence on the attractiveness of the townscape.
> (2) Calculation of the difference in attractiveness between the components = [weight].

The attractiveness of a townscape is created by the influence of its components.

The method of extracting weights in (2) is shown in the example of extracting the influence of Xn on the attractiveness of the townscape S. The component Xn is eliminated from townscape S to create [townscape S-Xn], and the difference in townscape attractiveness between [townscape S] and [townscape S-Xn] is extracted. The amount of change in attractiveness is caused by the component Xn and is considered to be the influence of the component Xn on the attractiveness of the townscape S. This makes it possible to calculate the influence of each townscape component on the attractiveness of the townscape.

Extraction of Major Townscape Components by SD Method and Multiple Regression Analysis

Here, we adopt the semantic differential (SD) method and multiple regression analysis method to conduct an impression survey.
First, we conduct the following process.
> 1) Awareness survey and numerical analysis using townscape photographs
> 2) Impression survey by simulation using 3D townscape images

Commercial streets in four locations of Tokyo (Ginza, Omotesando, Shimokitazawa, and Daikanyama) were selected for the analysis, as they are considered to be significantly influenced by existing values.
In (1), a questionnaire survey using townscape photographs was conducted on thirty-one

subjects. First, the attractiveness value of each townscape was quantified by using the SD method. The average value of the five items was set as the attractiveness value (Fig. 4).

Next, in the extraction of major components, we asked respondents to point out the components they felt were related to the Good and Wrong of each townscape (A-D) from the townscape photographs.

The sum of the number of comments on Good and Bad of each component extracted from the questionnaire is considered to be the component's influence on the current attractiveness of the townscape. These elements were selected as a result of the selection process.

The weighting of the elements was based on the data extracted from the previous awareness survey.
Attractiveness value of the townscape = Objective variable
The weighting of the elements was based on the data extracted from the awareness survey and was determined through multiple regression analysis.
> The weighting of the components was determined by multiple regression analysis.
> (The explanatory variables are analyzed separately for the "building element group" and the "street element group.")

The results were obtained by multiple regression analysis.
Multiple correlation coefficient R indicates the accuracy of the analysis.
The weights of the components are calculated based on multiple regression coefficients for each component, which expresses the weight of the explanation.
Weight = $|$ multiple regression coefficient $|$ × multiple correlation coefficient R× correction coefficient (10)

Simulation images with and without the components were created based on these data, and the process was repeated to survey the subjects' impressions. The following are examples of the exchange test in Ginza and Shimokitazawa. The reconstruction of the townscape using the exchange test is expected to be an effective tool in the consensus-building process to explore the direction of townscape restoration.

Example of townscape exchange test in Nihonbashi Ningyocho

This section presents a case study of a design code (rules for townscape formation) for the presence or absence of various townscape elements, referring to the townscape regulation, the "Landscape Law" enacted in 2004. We conducted a townscape simulation through a townscape exchange test to promote awareness among the administration and citizens and to create basis for establishing townscape design guidelines (rules for regulating the exterior design of buildings). The "Nihonbashi Ningyocho" was selected as the target site and as a result, building height, color, electric poles, and signage were identified as strong elements that influence the attractiveness of the townscape to create a "Ningyocho-like" atmosphere (details omitted).

The structure of "local character," which is expressed through a complex interplay of various townscape elements, was analyzed in a more in-depth manner in terms of townscape components. Although we can refer to the reproducibility of the townscape up to a certain level, there is a limitation in treating the visual components alone, because the actual regional townscape has other environmental elements, such as "various livelihoods," "sense of life," "smell," and "sound" in the region, as it is called "Seikatsukei" (living townscape). It is important to further layer other analytical methods on the results of the analysis here.

Chapter 2 — 2 差異による景観分析
1. 差異面による街並み景観のシミュレーション

概要

　一章（13頁）では、記号論における差異に注目し、差異面という2つの空間の質の差を表すグラフを境界上に立てるツールの開発について述べた。ここでは、我々の知覚において、それらの差異がどのように作用するのか。また、空間内における差異の分布が一目でわかるように変換されたグラフから何を読み取れるかについて、基本的な解説を行う。具体的には、伝統的な街並みや近代的な街並みを、差異面による分析を通して、街並みの特質を明らかにすることを試みる。差異面の立ち上げ方は、対数関数を使う方法、あるいはベクトル関数を使う方法などが検討されるが、ここではベクトル関数による方法を提示する。

分析対象の抽象化

　複雑な事物の集合体である街並みを対象として、その構成要素の関係性について分析する場合、大きさ、形、材料、テクスチュア、色などの構成要素に備わっているさまざまな性質を取り扱う必要があるが、それらは多層な次元にわたっており、同時に扱っても明解な分析結果を得ることは困難である。そこで、まず街並みを抽象化することによって、その構成原理を明らかにすることを試みる。この場合、街並みの立面を取り扱うことが通常であるが、その立面は直立投影面上に街並みを正投影したものであり、建物の出張りや引込みが抽象化されて、それ自体がより単純化されるため、二次元的面構成としてとらえることができる。建築のファサードは、建物の構造が常に垂直方向に働く重力に対抗するものであったため、垂直・水平線のみによって構成されているところに特徴がある（Fig.1）。

（Fig.1）抽象化された街並みの立面
街並みを二次元的な立面図に表現して分析対象とする

構成要素の形の差異を視覚化するための方法

　デザイン論では、一般的に矩形の縦と横の比率には美しいと感じるおおよそ一定の割合があり、その比率から大きく離れてしまっているものは、美しくないとされてきた。これを数理的な法則によって規定したものが"黄金比矩形"や"ルート矩形"である。我々は実際の設計行為の際に、無意識にその矩形の形状を表すような要素を引き出しており、この要素の一つを対角線としてとらえてきた面がある。ロシア出身の画家ワシリー・カンディンスキー（1844～1944年）は、対角線を"緊張

測定線"として、矩形のもっている内面的緊張を示す重要な要素であるとした。そして、そのズレによって、矩形がさまざまな性質を帯びてくることを述べている。　　実際に、我々は矩形の形状やプロポーションを比較する時にそれぞれの対角線の存在を感じ取っているのであろう。それゆえに、矩形のさまざまな状態をあらわす要素として、対角線というパラメーターは有効であると考えられる。さらに考察を進めると、矩形のさまざま

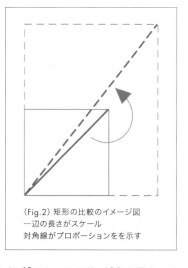

（Fig.2）矩形の比較のイメージ図
一辺の長さがスケール
対角線がプロポーションをを示す

な状態とは即ち、「縦と横の比」および「スケールの比」が各々異なっているということに気づく（Fig.2）。矩形の対角線の長さを比較することは、即ちそのスケールを比較することに等しく、対角線の角度を比較することは、縦と横の長さのプロポーションを比較することに等しいと考えられるため、街並みの形態を分析するための大きな拠り所となると考えられる（Fig.3）。以上の考察から、ここでの分析では矩形の形のさまざまな状態を表す数式として、対角線の長さと角度を用いる。

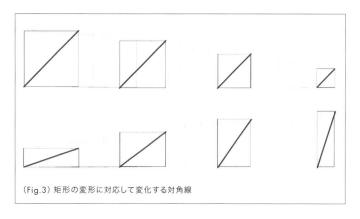

（Fig.3）矩形の変形に対応して変化する対角線

「差異面生成システム」を使用した街並みの分析手順

　本分析では、矩形のさまざまな状態を表す数式として対角線の長さと角度を用い、これらを比較してその差異を視覚化することによって、街並みの構成原理を解き明かすことを試みるが、街並みを構成する要素の数は非常に多数であり、それぞれの比較を手作業で行うと非常に多くの労力を費やすため、コンピュータ・プログラムによって半自動的に差異面を生成させるシステムを開発した。以下が、コンピュータ・プログラム「差異面生成システム」を使用した分析の手順である。

【手順1】抽象化した街並みの立面を CAD によって作図する（Fig.4）

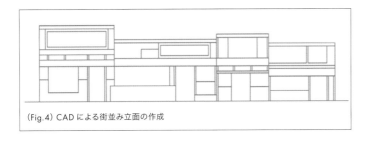

（Fig.4）CAD による街並み立面の作成

【手順2】T字にならないコーナーをもつ線分については、そのまま次の線分にぶつかるまで延長する。これによってすべてが矩形に分割される（Fig.5）

（Fig.5）矩形による分析対象図の作成

【手順3】それぞれの矩形について対角線を作図する（Fig.6）

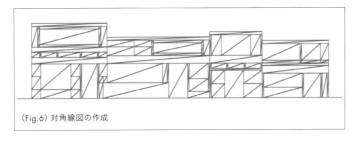

（Fig.6）対角線図の作成

【手順4】以上の手順で作成した CAD データを取り込み、コンピュータによる点と線分の分析情報を画面に出力させる。

　以上の手順に従って、実際に街並みの分析を試みた。この方法による街並みの分析により、基本的に矩形で構成されている建物の立面は、どのような組み合わせの構成であっても、同じ方法論で比較できるようになった。（オーガニックな建物形態については別の方法が求められる。）

差異面による街並み分析の実践

　街並みのような複雑な構成要素の集合体について分析を行う場合、差異面の分布状態は明瞭なものではなく、非常に複雑なものになる。しかし、実際にいくつかの特徴的な街並みについて分析し、その結果から複雑さの中に潜む何らかの構成秩序を見出すことができれば、差異面による街並みの分析方法の有効性を示すことができる。ここでは、特徴的と思われる三重県の関町（日本の伝統木造による街並み）、ザルツブルグ市ゲトラーデ通り（西欧の石造りによる伝統的な街並み）、東京都丸の内（近代的な箱型建築による街並み）についての分析を試みる。

1）日本の伝統木造における街並みの分析

（Photo 1）関町の街並み

　三重県鈴鹿市関町は鈴鹿山脈の東の山裾にあり、昔は鈴鹿の関が置かれていた交通の要所であった。東海道有数の難所であった鈴鹿峠から鈴鹿川の谷間を抜け、伊勢平野に出た地点に開かれた宿場町である。今でも、旧東海道に沿って東西に約1,800mにわたり、伝統的な町家が軒を連ねており、現在も近代以前の空間特性のなごりをとどめている（Photo1）（Fig.7）。この変化に富みながら統一感のある街並みの構成秩序を、差異面による分析によって明らかにする。

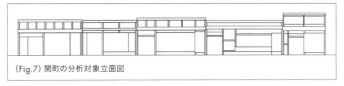

（Fig.7）関町の分析対象立面図

　差異面分布図を見ると、壁面にあらわれる柱型という縦に細長い構成要素が、軒という横に細長い構成要素に接している部分に高くて細長い差異面が立ち上がっている。これがいくつも連続しているのが関町の差異面分布状態の特徴である（Fig.7）。このような差異面の分布状態

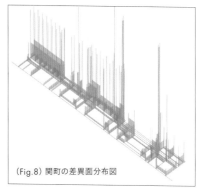

（Fig.8）関町の差異面分布図

は、水平性の強い要素（軒）と垂直性の強い要素（柱）の組み合わせによって生じている。関町の伝統的な町家は、シンメトリーではなく同じような要素によって構成されながら、みな異なったつくりをしているために、多少ばらつきながらも高くて細長い特徴的な差異面が連続するといった分布状態になっており、これが関町の伝統的な街並みにリズムを与えていると考えられる。差異面分布図を部分的にみると差異面がギザギザに立ち上がっているために一見まとまりがないように見えるが、全体を見ると差異面の立ち上がり方に類似性がみられるために統一感のある街並みになっている。また、高くて細長い特徴的な差異面によって、明らかに軒の水平的連続性が強調されている。この分析から、多様な差異面が立ち上がりながらも、建物や街並みの局所部分において軒と柱の関係によって生じる差異面の分布状態が類似しているということが、伝統的木造街並みを変化に富みながらも統一感のある街並みにさせているということが分かる。

2）西欧の石造りによる伝統的な街並みの分析

　ゲトラーデ通りはオーストリアとドイツの国境に位置するホエン・ザルツブルグ城の城下町の中にあり、現在も中世の雰囲気を漂わせる歴史的な街並みを残す通りである。幅5mほどの街路の両側には、それ

（Photo 2）ゲトラーデ通りの街並み

（Fig.9）ゲトラーデ通りの分析対象立面図

ぞれに個性のある石造りの建物が隙間なく連なっており、全体的に統一感のある魅力的な街並みを形成している。ここでは差異面による分析によって、何がゲトラーデ通りの街並みに統一感やリズムを与えているのかを明らかにする。（Photo2）（Fig.9）

　差異面分布図を見ると、一層部分の開口のスケールとその上部の小窓部分のスケールが明瞭に分節されていることが分かる（Fig.10）。一層部分では差異面がほとんど立ち上がっていないのに対して、上部では差異面の規則的な分布がみられ、全体をひとつのまとまりとして認識することができる。この上部に見られる規則的な差異面は、壁面に規則的に開けられた開口部によって

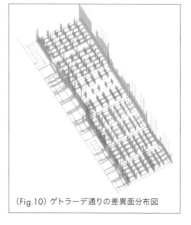

（Fig.10）ゲトラーデ通りの差異面分布図

立ち上げられているが、これが通りの街並みに統一感や連続性を与えている。また、街並みを構成している建物のそれぞれの開口部が少しずつ違った形をしているために、差異面の規則的な立ち上がり方の中に微妙な差異が生じており、この微妙な違いが街並みに心地よいリズムを与えている。また、街並みの上端の装飾部コーニスに水平に連なる高い差異面がみられるため、スカイラインが揃っているという印象をより強く受ける。このような差異面の分布状態はゲトラーデ通りだけにみられる特徴ではなく、西欧の組積造建築による街並みの大多数にみられる特徴といってよいであろう。この分析から、ゲトラーデ通りの街並みにみられる統一感やリズムは、平面的に連続する壁面とそ

こに規則的に開けられた開口部の組み合わせによって生じているということが分かった。

3）近代的な箱型建築における街並みの分析

　明治維新時に、東京の丸の内地区には新政府の中枢機関が置かれたが、明治22年（1889年）の「東京市区改正設計」によって経済地区として整備されることになり、翌年に三菱社が国策に協力する形で日本初のオフィスビル街の整備を始めた。この時、丸の内の街並みは西欧風の佇まいであったために「一丁ロンドン」と呼ばれていた。現在の近代的大型ビルの連なる街並みは、昭和25年（1950年）に勃発した朝鮮戦争の特需景気をきっかけとした高度成長期の中で「丸ノ内総合改造計画」などの大規模な再開発計画によって形成されたもので、残念ながら明治時代の建物のほとんどは取り壊されてしまい、昔の面影はない（Photo3）（Fig.11）。

（Photo3）丸の内の街並み

　差異面分布図を見ると、右側の建物の部分に非常に高い差異面が立ち上がっているのが分かる（Fig.12）。これは各層にある水平性の強い要素によって立ち上げられた差異面であるが、周囲に対して高いというわけではなく、すべてが高いという差異面の分布状態であるために、"図"というよりも全体としての"地"としての性格のほうが強いと考えられる。また、差異面が全く同じくり返しであるために、リズムが単調であるといえるだろう。この建物は全体的にみるとやはり異質である。中央の建物に関しては、一層と二層の間に高い差異面が見られるが、これは社名の入っている部分で、他の要素に対して強調されていると見ることが出来る。二層以上は一様に低い差異面の分布状態で、変化に乏しい立面になっており、方

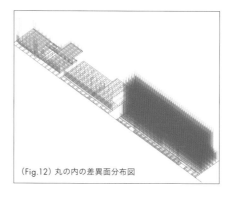

（Fig.12）丸の内の差異面分布図

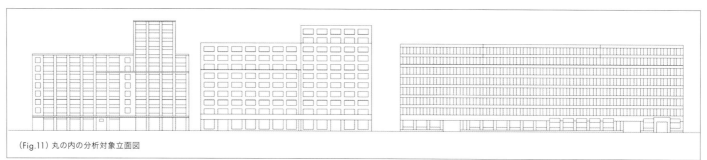

（Fig.11）丸の内の分析対象立面図

向性もない。左側の建物に関しては、縦の柱型によって立ち上げられた差異面がほぼ同じ高さで連なっていることから、垂直性の強いデザインであるといえる。全体的な街並みとして見ると、差異面の分布状態としてはあまり共通性が見られないが、それぞれの建物における差異面の分布状態が均質であることは明白である。街並みに感じた唯一の統一感は、要素の構成というよりは、色彩やスカイラインが揃っていることなどの他の要因によるところが大きいと考えられる。

これら、風土も歴史も異なった三地区の街並みを差異面で分析したが、近代に入り、画一的な工業製品を用いた合理的な工法により、早く安く建物を建てようとした結果、近代以降の街並みは非常に単調になり、どこの場所でも量産できるものになっている。

以下は、伝統的街並みと近代的街並みの混在による差異面の立ち上がりの類型を示している。立ち上げりの乱雑さにより、私たちの視覚も混とん状態として街並みを把握していることが予測される。

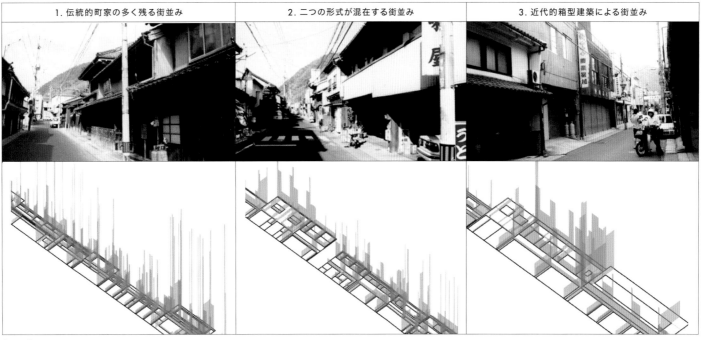

（Fig.13）

以下は、伝統的街並みの連続性が異質な建物や空間で分断されている類型を示している。

規則的な差異面の連続性が、突然断ち切られ、我々の視覚における水平連続性の認識が実際に分断されていることが分かる。

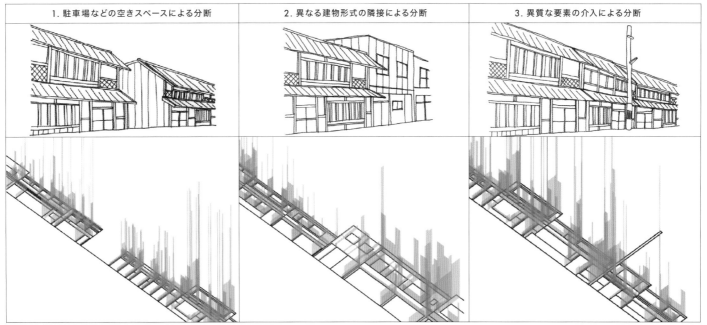

（Fig.14）街並みを分断する差異面の分布状態の類型化

Morphological Analysis of Townscape by Differential Plane

SUMMARY

In Chapter 1 (p.13), we focused on differences in semiotics and described the development of a tool called the "differential plane," which represents quality differences between two spaces, illustrated through a graph at the boundary. Furthermore, we provide a basic explanation regarding how to read the transformed graph so that the distribution of differences within a space can be seen at a glance. In this section, we attempt to identify the characteristics of traditional and modern townscapes through analysis using the differential plane tool, and here, we use the vector function as the start-up method instead of a logarithmic function.

Abstraction of the object of analysis

When analyzing the relationships among the components of a townscape, which is a complex collection of objects that contains various properties, such as size, shape, material, texture, and color, and cover multiple dimensions, it is difficult to obtain clear results when dealing with them simultaneously. Therefore, we will attempt to clarify the compositional principles of the townscape by first abstracting it into a two-dimensional surface composition.

Methods for visualizing differences in the shapes of the components

In design theory, a certain ratio perceived as beautiful between the height and width of a rectangle are defined in mathematical rules as the "golden ratio rectangle" or "root rectangle." In our design process, we unconsciously draw out elements that represent the shape of a rectangle, one of which is the diagonal. The Russian painter Wassily Kandinsky (1844-1944) considered the diagonal to be a "tension measuring line," an important element that indicates the internal tension of the rectangle and takes on various properties depending on its misalignment. Further investigation reveals that various states of a rectangle can be identified by the "ratio of length to width" and the "ratio of scale." Comparing the length of the diagonals of a rectangle is equivalent to comparing its scale, while the angle resembles the proportions of the length and width, and these can be used as the basis for analyzing the form of a townscape. Based on the above considerations, the analysis in this section applies the length and angle of diagonals as mathematical expression to depict various states of rectangular shape.

Procedure for townscape analysis using the "Differential Plane Generation System"

In this analysis, the lengths and angles of diagonals of rectangles are used to compare and visualize the differences in order to elucidate the principles of townscape composition. We thus developed a system to generate differential planes semi-automatically using a computer program. Steps for "Differential Plane Generation System" analysis on the computer program are as follows.

Step 1: Draw an elevation of the abstracted townscape using CAD. (Fig.4)

Step 2: Extend the line segments with corners that do not form a T-shape until they meet the next line segment. This divides everything into rectangles. (Fig.5)

Step 3: For each rectangle, draw a diagonal line. (Fig.6)

Step 4: CAD data created in the above steps is imported, and the computer outputs the point-and-line segment analysis information to the screen.

By analyzing the townscape using this method, building elevations that are composed of rectangles can now be compared using the same methodology.

Townscape analysis using differential planes

When analyzing a complex set of components such as a townscape, the state of distribution of differential plane is not clear and is very complex. However, if we can analyze and extract some townscape characteristics and compositional order hidden within the complexity from the results, we can show the effectiveness of this method. Here, we attempt to analyze three townscapes with different characteristics: Sekimachi, with a traditional wooden Japanese structure, Getrade Street in Salzburg, with a Western stone structure, and Marunouchi in Tokyo, with a modern box-shaped structure.

1. Analysis of townscape in traditional Japanese wooden construction

Sekimachi, Suzuka City, Mie Prefecture, is located at the eastern foothills of the Suzuka Mountains and was once an important transportation hub where the Suzuka-no-seki barrier was located. Traditional townhouses still stand side by side along the old Tokaido Highway for approximately 1,800 meters from east to west, retaining the spatial characteristics of the pre-modern era. This study clarifies the order in which this varied yet unified townscape is organized by analyzing the differential planes.

Differential Plane Distribution Map and Discussion

The differential plane distribution map shows that the long and narrow columnar elements on the walls are high and narrow where they are in contact with the long and narrow horizontal elements called eaves. The continuous distribution of these surfaces is a characteristic of the distribution state seen from the differential planes in Sekimachi due to the combination of strongly horizontal elements (eaves) and strongly vertical elements (columns). The traditional townhouses in Sekimachi are not symmetrical, but are composed of similar elements. Still, they are all built differently, resulting in a continuous distribution of high and slender features of differential planes that vary slightly. Looking partially at the distribution map, the jaggedly extruded differential planes seem to lack cohesiveness. However, the similarity seen in the patterns of extrusions when viewed as a whole creates a sense of unity in the townscape. This analysis shows that the similarity in the distribution of different surfaces caused by the relationship between eaves and columns in local areas of the buildings and streetscape gives the traditional wooden townscape a sense of unity despite its rich variety.

2. Analysis of a traditional townscape made of stone in Western Europe

Getrade Street is located in the castle district of Hohen Salzburg Castle on the border between Austria and Germany. This street retains a historical townscape with a medieval atmosphere. Stone buildings, each with their own unique character, stand on both sides of the five-meter-wide street, forming an attractive townscape with a sense of unity. Here, we will analyze what gives the streetscape of Getrade Street a sense of unity and rhythm using differential planes. (Photo 2)

Differential Plane Distribution Map and Discussion

The differential plane distribution map shows that the scale of the opening in the first layer is clearly divided from the scale of the small window above the opening. There are almost no extrusions of differential planes, while the upper part shows a regular distribution, which can be recognized as a single unit. The consistency seen in the upper part is due to the regular openings of the walls, which give a sense of unity and continuity to the streetscape, whilst the slight differences in the shapes of building openings creates a streetscape with a pleasant rhythm, which is represented in the subtle dissimilarities within the regular extrusions of the differential planes. This analysis indicates that the sense of unity and rhythm in the streetscape of Götrade Street is the result of the combination of flat, continuous wall surfaces and regular openings.

3. Analysis of the streetscape in modern box-shaped architecture

At the time of the Meiji Restoration, the Marunouchi district of Tokyo was home to the central government offices of the new government, and was to be developed as an economic district according to the "Tokyo City District Revision Design" of 1889. The following year, the Mitsubishi Company began to develop Japan's first office building district in cooperation with national policy. At this time, Marunouchi was known as "Iccho London" because of its Western-style appearance. Unfortunately, most of the buildings from the Meiji era have been demolished, and no trace of the old days remains. (Photo3, 4)

Differential Plane Distribution Map and Discussion

The differential plane distribution map shows that the buildings on the right side are higher. This is caused by elements with strong horizontal characteristics in each layer, not necessarily because they are higher in relation to its surroundings. Due to its distribution state with all showing high, it has a stronger character as a "ground" rather than a "figure." Furthermore, despite the slight commonality seen in the distribution of differential planes when looking at the overall townscape, it is clear that the distribution of each building is homogeneous. This implies that the sense of unity in the townscape is not influenced by the composition of the elements, but rather by other factors such as the uniformity of the colors and skylines.

In contrast to the differences in the townscapes of these three districts with varying climates and histories, the townscapes of the modern era have become very monotonous and mass-produced in every town as a result of the attempt to build buildings quickly and cheaply using rational construction methods with uniform industrial products. Fig. 12 presents the types of differential plane extrusions due to the mixture of traditional and modern townscapes. Their cluttered nature suggests that our visual perception of the townscape is also in a state of confusion.

Fig.13 presents a traditional townscape, the continuity of which is broken up by heterogeneous buildings and in-between spaces.

Fig.1   Elevation of abstracted streetscape
Fig.2   Imagery of rectangle comparison
Fig.3   Diagonal lines changing in response to rectangle deformation
Fig.4   Creation of street elevation by CAD
Fig.5   Creation of the diagram to be analyzed using rectangles
Fig.6   Diagonal diagram
Photo1   Sekimachi streetscape
Fig.7   Elevation of Sekimachi
Fig.8   Differential Plane distribution map of Sekimachi
Photo2   Streetscape of Getrade Street
Fig.9   Elevation of Getrade Street
Fig.10   Differential Plane distribution map of Getrade Street
Photo3   Streetscape of Marunouchi
Fig.11   Analyzed elevation of Marunouchi
Fig.12   Distribution of differential planes in Marunouchi
Fig.13 and 14   Typology of the distribution of differential planes dividing the streetscape

概要

　差異面という分析ツールを用い、一般的に色の差と言われるものをグラフ化することで、色差の境界における差異のあらわれ方を具体的に検討する。一般に、色弱あるいは色盲の人々は、さまざまな色の特質の差異を把握できないといわれている。したがって、補色のように明らかに色の差が違う領域や、同じトーンで連続的に変わる領域などの間に、差異のグラフがどのように示されるのかという視点も興味深い。具体的には、差異度グラフを用いて、さまざまな街並みにおいてケーススタディーを行い、「図」となる要素を抽出し、それらの質や密度の観点からそのまちの特徴を見出す。また、それを通じて残すべき景観要素と改善すべき景観要素を検証し、今後の街並み整備のあり方やサイン計画の指針を模索する。

色差の視覚化

　ものの知覚において、色彩は重要な要素で、私たちは色の差によって、ものを識別している。つまり、私たちがものを認識するためには、背景との色彩の差（色差）が必要であり、それらの差異により、「地」から分節して「図」を切り取って認識しているのである。特

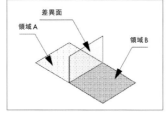

（Fig.1）色差異面のイメージ

に知覚認識においては、相対的な差異の度合いが重要であり、差異が大きいほど強い「図」として認識される。したがって、どのような要素を強く知覚しているのかという街並みの認識構造の一端を解明するには、ある要素がどの程度ほかの要素と異なっているのかという差異の大きさが重要になり、それを分かりやすく示すことが必要となる。それを表現するための方法として差異面による分析を行う。差異面とは、隣接する2要素間における質が、ある次元においてどれくらい異なっているのかを物理量に変換し、隣接する2要素の接線に高さのある面を立ち上げ、視覚化したものである（Fig.1）。ここでは色彩の差による差異面、つまり"色差異面"によって、隣接する要素間における色彩の差の関係性を視覚化する。

色差（⊿E）について

　色彩には、色味を示す色相、明るさを示す明度、鮮やかさを示す彩度の3つがある。それぞれの差異を色相差、明度差、彩度差で示すことができ、目的によっては有効な手段であるが、多様な色彩により構成されている街並みを分析するうえでは、色相、明度、彩度を総合的に考える必要がある。ここでは、三要素を1つに統合して算出する色差を用いる。

　色差とは指標上における二色間の距離のことをいうが、⊿Eで示す。⊿Eを算出する方法としてCIEL*a*b*表色系による方法、CIEL*u*v*表色系による方法、アダムス‐ニッカーソン色空間による方法、ハンター色空間による方法、CMC(l:c)色差式による方法などがあるが、ここでは、国際照明委員会（CIE）で規格化されている方式CIEL*a*b*表色系による方法を用いることとする。

色表示値の相互変換について

　一般に、コンピューターにおける色彩表示にはモニター RGB（sRGB）が用いられている。したがって、色差の算出にあたっては、モニター RGB（sRGB）からCIE L*a*b*表色系に変換する必要がある。変換は以下の手順に従って行う。

コンピューターにおける色彩表示値の変換

モニター RGB（sRGB）⇒ リニア RGB ⇒
⇒ CIE XYZ 表色系 ⇒ CIEL*a*b* 表色系

　2色間の色差⊿Eにおける色相差、明度差、彩度差、トーンによる比較を（Fig.2）に示す。

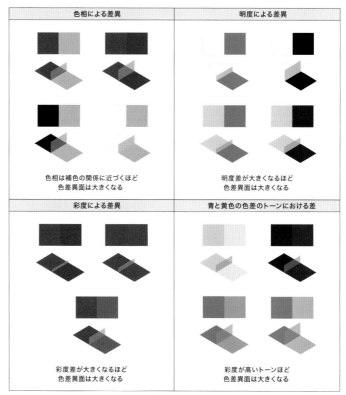

（Fig.2）2色間の色差⊿Eの色相差、明度差、彩度差、トーンによる比較

まちの色差を差異面で見える化する

1）色差異度グラフにより、街並みにおいて強く知覚される図的要素の抽出を行い、さらに色差の強さから図的性質の強度について考察を行う（Fig.3）。

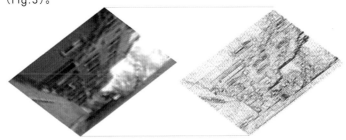

（Fig.3）ボストンの街並みの色差に関する分析事例

2）色相／トーングラフにより、街並みがどのような色彩によって構成されているかを考察し、色彩が知覚構造にもたらす影響について検証を行う（Fig.4）。

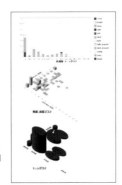

（Fig.4）ボストンの街並みの色相とトーンによる分析事例

3）色差異度グラフの立ち方により、集合体を構成する各要素間の色差異面の分布状態を示す（Fig.5）。

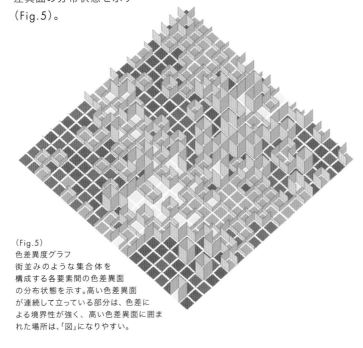

（Fig.5）
色差異度グラフ
街並みのような集合体を構成する各要素間の色差異面の分布状態を示す。高い色差異面が連続して立っている部分は、色差による境界性が強く、高い色差異面に囲まれた場所は、「図」になりやすい。

　主に高い色差異面が連続して立っている部分や色差異面の密度に大きな差が生じているところを境界として、知覚的には「図」と「地」に分化されている。
　（Fig.6）は色差異面の立ち上がりによる図面の類型化を図ったものである。

線的な図		
輪郭型	密着差型	
	少色相型	多色相型
A	B	C

線的な図				
輪郭型	密着差型			
	図の密度が高い		地の密度が高い	
	少色相型	多色相型	少色相型	多色相型
D	E	F	G	H

（Fig.6）色差異面の立ち上がり方の類型
主に高い色差異面が連続して立っている部分や色差異面の密度に大きな差が生じているところを境界として「図」と「地」に分化されている

　以下では、ベネチア（イタリア）、高山（岐阜）、渋谷（東京）、新宿（東京）各地区の景観の色差異面グラフによる分析事例を示す。

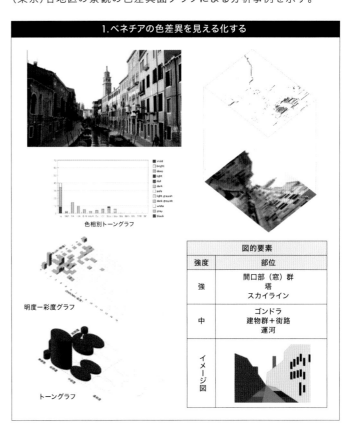

1.ベネチアの色差異を見える化する

色相別トーングラフ

明度一彩度グラフ

トーングラフ

図的要素	
強度	部位
強	開口部（窓）群 塔 スカイライン
中	ゴンドラ 建物群＋街路 運河
イメージ図	

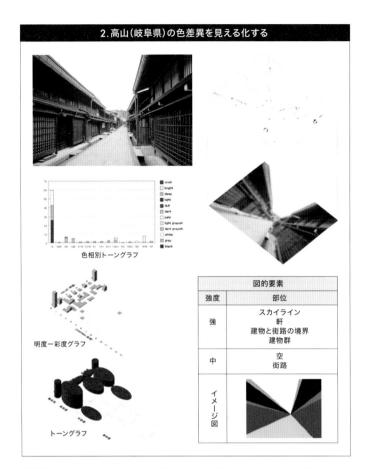

2.高山(岐阜県)の色差異を見える化する

色相別トーングラフ

明度一彩度グラフ

トーングラフ

図的要素	
強度	部位
強	スカイライン 軒 建物と街路の境界 建物群
中	空 街路
イメージ図	

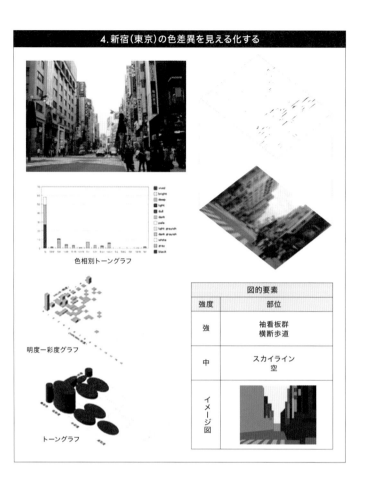

4.新宿(東京)の色差異を見える化する

色相別トーングラフ

明度一彩度グラフ

トーングラフ

図的要素	
強度	部位
強	袖看板群 横断歩道
中	スカイライン 空
イメージ図	

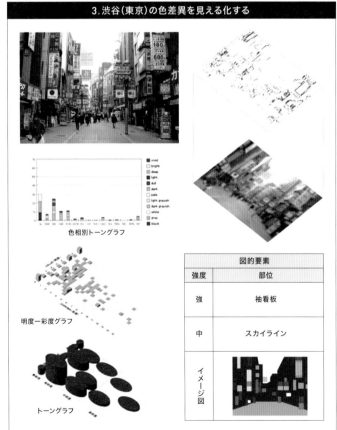

3.渋谷(東京)の色差異を見える化する

色相別トーングラフ

明度一彩度グラフ

トーングラフ

図的要素	
強度	部位
強	袖看板
中	スカイライン
イメージ図	

分析による考察

　ベネチアや高山のような伝統的街並みにおいては、色相も安定し、トーンのブレも少ない。このような地域では、古くから地場の建築材料を用い、予算をかけないローコストな建物を旨としたので、色彩的な差異も少ない類似色相型となっている。「図」としてのまとまりもあり、記憶に残りやすい景観の知覚認識となっている。一方、渋谷や新宿では、商業地域のため、客の目を引く彩度の高い看板が多く、多色相型となっている。また、色差異面が乱雑に立っているため、「図」としての認識が弱く、記憶に残りにくい景観の知覚認識となっている。

　これらの分析で得られた知見を、現代都市景観における色彩規制や伝統的景観の修復などに応用していくことができる(Fig.6)。

（Fig.7）街並みの知覚構造と色彩による類型化

SIMULATION

Analysis of Color Difference by Differential Plane

SUMMARY

Using an analytical tool called a differential plane, we will examine how differences appear to our eyes at the boundaries of color areas by graphing the level of differences. In general, people who are colorblind are unable to perceive differences in various color qualities. Therefore, it is also important to consider how the graph of differences is shown between areas of clearly different colors, such as complementary colors, and areas of gradational change in the same tone. Specifically, using the differential plane graph, case studies are conducted in various townscapes to extract "figure" elements, and the characteristics of the town will be found in terms of their townscape quality and density. Through this process, we examine the townscape elements that should be preserved and those that should be improved and seek guidelines for future townscape development and signage planning.

Visualization of color differences

Color is an important element in our perception, and we identify objects by their color differences. In other words, we need contrasts of color differences in relation to the background in order to recognize objects, and these differences allow us to separate and recognize the "figure" from the "ground." Especially in perceptual recognition, the relative degree of difference is important: the greater the difference, the stronger the "figure" is perceived. Therefore, in order to clarify the cognitive structure of the townscape, it is important to determine the degree of difference between an element and other elements to be understood easily. The differential plane is used to express this analysis, visualizing how much the quality between two adjacent elements differs in a certain dimension, describing a graphic plane with physical height at the boundary line of the two adjacent elements. Here, we visualize the relationship between the differences in color between adjacent elements by means of a "color differential plane" (Fig. 1).

Color Difference ($\triangle$E)

There are three elements of color: hue, which the color; lightness, which indicates brightness; and saturation, which indicates vividness. The differences can be indicated by hue difference, brightness difference, and saturation difference, which are effective tools for specific purposes. However, in analyzing a townscape composed of various colors, it is necessary to consider these three elements in a comprehensive manner. Here, we use color difference, which is calculated by integrating the three elements into one.

Color difference refers to the distance between two colors on an index and is indicated by $\triangle$E. There are several methods for calculating $\triangle$E, such as the CIEL*a*b* color system, the CIE L*u*v* color system, the Adams-Nickerson color space, the Hunter color space, and the CMC (l:c) color difference formula, but here we use the method based on the CIEL*a*b* color system standardized by the International Commission on Illumination (CIE).

Mutual Conversion of Color Indication Values

Generally, monitor RGB (sRGB) is used for color displays on computers. Therefore, it is necessary to convert from monitor RGB (sRGB) to the CIE L*a*b* color system when calculating color differences. The conversion is performed according to the following procedure.
Conversion of color display values on a computer

Monitor RGB (sRGB) $\Rightarrow$ Linear RGB $\Rightarrow$ CIE XYZ color system $\Rightarrow$ CIEL*a*b* color system

As a comparison by hue, lightness, saturation, and tone difference in color, $\triangle$E between two colors is shown in (Fig. 2).

Visualizing the color difference of a town using differential planes

(1) Using the color difference plane, we extract the outstandingly perceived graphic elements in the townscape, and furthermore, from the level of the color difference, we discuss the "figure" strength.

(2) By using hue/tone graphs, we examine what kind of colors compose the streetscape and verify the influence of color on our perceptual structure.

(3) The color differentiation graph shows the state of distribution of color differential planes among the elements that compose the aggregate.

Fig 5 presents the distribution of color-differential planes among the elements that make up an aggregate such as a townscape. Areas in which high color-differential planes stand consecutively have strong boundary characteristics due to color differences, and areas surrounded by high color-differential planes tend to be "figures."

Mainly, areas in which high color-differential planes stand in succession or where large differences in the density of color-differential planes occur are strongly precepted onto "figure" and "ground."

Fig. 6 presents the typology of drawings based on the rise of color-differentiated surfaces.

(Fig.6) Typology of color-differential planes: The "figure" and "ground" are mainly differentiated at the boundaries of areas in which high color differential planes stand in succession and where large differences in the density of color differential planes are observed.

The following images are examples of analysis using color-differential plane graphs of landscapes in Venice (Italy), Takayama (Gifu), Shibuya (Tokyo), and Shinjuku (Tokyo).

Visualization of color differences in Venice (Italy)

Visualization of color difference in Takayama (Gifu Prefecture)

Visualization of color differences in Shibuya (Tokyo)

Visualization of color difference in Shinjuku (Tokyo)

Discussion through analysis

In traditional townscapes such as Venice and Takayama, hues are stable and tones are not blurred. In such areas, local building materials have been used for a long time and low-cost buildings with a limited budget have shaped the townscape, resulting in similar hue types with few differences. Hence, the "figure" is coherent and the perceptual recognition of the landscape is easy to remember. In contrast, in the commercial areas of Shibuya and Shinjuku, there are many highly saturated signboards that attract customers' attention, resulting in a multicolor hue type. In addition, because the color-differential planes stand in a cluttered manner, the perception of the townscape as a "figure" is weak and the memory of the townscape is difficult to sustain. The findings from these analyses can be applied to the regulation of color in modern urban townscapes and the restoration of traditional landscapes.

Fig. 1   Image of color differential plane
Fig.2   Comparison of color difference between two colors by hue difference, lightness difference, saturation difference, and tone
Fig. 3   Example of analysis on color differences in a Boston townscape
Fig. 4   An example of analysis of a Boston cityscape using hue and tone
Fig.5   Color difference plane graph
Fig. 6   Perceptual structure of townscape and typification by color

1. 街路空間の視覚シークエンスを見える化する

概要

一章（16頁）では、無自覚に「見える」視覚と、意識を持って「見る」視覚の情報量に違いについて述べた。ここでは、光源投射法による視覚範囲である「見える」視覚を扱い、仮想空間の中で人間を移動させ、「見える」視覚領域の変化を探る。また、この変化に対する心理的変化量を同時に観察し、視覚的シークエンスが心理に与える変化のメカニズムを探る。

ビジブルエリアの考え方

人の情報処理能力の中で最も利用されるものは視覚であり、ほとんどの場合、都市は「見る」ことを通じて理解され、人は限られた視野の中からさまざまな情報を得て行動する。ここでは人によって「見える」範囲または認識可能範囲をビジブルエリアとし、光源投射シミュレーション方式により、街路空間の視覚シークエンスにおけるビジブルエリアの分析を行う。参考資料などから、人の動作が見分けられる最大の距離を135mと設定し、人の立つ地点から半径135mの円以内を、人が街路空間を認識することができる可能な範囲（ビジブルエリア）として分析する。街路空間には、「見る」行為を誘引するさまざまなものが存在し、それが個人により違いがあるために、一定方向のみのビジブルエリアを分析する方法だけでは、実際に人が歩いている状態と大きな差が生じると考えられる。そこで、ビジブルエリアは人を中心に360度広がる領域として分析を行う（Fig.1）。

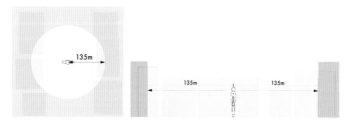

（Fig.1）ビジブルエリアの範囲（平面・立面）
人を中心に半径135m以内の円をビジブルエリアとし、周囲360度の全方向に関して分析を行う

観測地点について

街路空間における移動経路を15m間隔に分割し、15mおきに観測地点を設け、各観測地点のビジブルエリアを算出する。観測地点は街路空間における人の可視範囲を追及するために、歩道のない街路では街路の中心、歩道のある場所では歩道の中心とする。この距離単位は、成人男性が10秒間に歩く平均距離であり、近景の範囲内での移動であるため、ヒューマンスケールでの考察が可能となると考えられる。

ビジブルエリアと視覚シークエンス

光源投射法によりビジブルエリアの面積の変化を数値的に算出し、街路空間のシークエンスについて分析を行う。基本的には、人を点光源に置き換えることでビジブルエリアをモデル化し、まず人が視覚により空間情報を得ることが可能な範囲として示す。そして、人が移動するときのビジブルエリアの変化を視覚シークエンスとして捉える（Fig.2）。

光源投射シミュレーション方式によるビジブルエリアの算出

手順1	2次元地図の作成	
地図データをもとに、建物の配置や街路の形状などをもとに、2次元地図を作成する。建物の高さなども予め調査しておく。		
手順2	**観測地点の設定**	
観測地点を設定する。歩道がある街路では観測地点を歩道の中心、歩道のない街路では街路の中心とする。×が観測地点である。		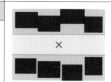
手順3	**3次元へ変換**	
手順1で作成した2次元地図に建物や塀などの視線を遮るものの高さの情報を与え、3次元地図に変換し、光を投射する際に、透過しないようにする。		
手順4	**光の投射**	
観測地点に点光源から投射された光が当たる領域がビジブルエリアとなる。視覚自体は距離により減衰するので、輝度のグラデーションで示す。		
手順5	**2次元地図の作成**	
手順4で視覚化されたビジブルエリアは無限に広がるので、半径135mの円を描き、内側のビジブルエリアの面積を求める。		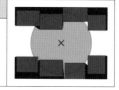

（Fig.2）ビジブルエリアを算出するためのプロセス
実際の建物の配置を3次元画像に立ち上げ、観察地点から投射した光によるビジブルエリアの面積をシミュレーションの根拠とする

ビジブルエリアの分析

光源投射シミュレーション方式により得られるビジブルエリアの情報は1）ビジブルエリアの形態と2）ビジブルエリアの面積量の2つである。

ビジブルエリアの形態は距離と大きさを持ったパターン情報、ビジブルエリアの面積量はスカラー情報であるといえる。ここでは、これらの情報について以下の分析を行う。

1）観測地点におけるビジブルエリアの形態については、基本的にはリニアーな街路に沿った「見える」景観に加え、T字路や交差点において移動方向とは垂直の方向への視覚が開けるときに大きな形態の変化が起こっている、ビジブルエリアの枝分かれの状況や枝の長さにその街路の特徴が良く現れるが、特に湾曲した街路では、ビジブルエリアの大きさが135mまで届かず、囲まれ感のある景観が生まれていることが分かる（Fig.3）。

2）各観測地点におけるビジブルエリアの面積とその変化量については、人の視覚構造が近くにある対象と遠くにある対象ではその解像度が異なり、光の減衰に似たグラデーション的減少をすることを考慮し、距離に応じた減衰係数をかけたものをビジブル面積としてカウントとすることとする。まず、ビジブルエリアを同心円状に10分割し、内側からエリア1～10とする（Fig.4）。

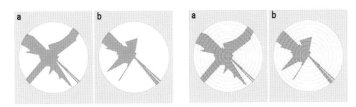

（Fig.3）観察点におけるビジブルエリアの表示
各地点の異なった形状を記録し、同心円状にマッピングする

（Fig.4）同心円内のエリア別の面積計算
ビジブルエリアの面積を10個のエリア別に算出する

次に、10分割したビジブルエリアに観測地点からの遠近を考慮し、異なった係数をかけ比較する。人の目は光情報を集めることで見ることが可能となるため、係数は、光の減衰度＝1／距離の二乗（光源に垂直な面における照度の減衰）を用いることにした。観測地点からエリアの中心までを距離とし、その距離と係数は次の通りである（Fig.5）。

	距離	係数
エリア1	675cm	$1/(675)^2$
エリア2	2025cm	$1/(2025)^2$
エリア3	3375cm	$1/(3375)^2$
エリア4	4725cm	$1/(4725)^2$
エリア5	6075cm	$1/(6075)^2$
エリア6	7425cm	$1/(7425)^2$
エリア7	8775cm	$1/(8775)^2$
エリア8	10125cm	$1/(10125)^2$
エリア9	11475cm	$1/(11475)^2$
エリア10	12825cm	$1/(12825)^2$

（Fig.5）視覚の減衰度を光の減衰度と同等と想定した表
分割されたエリア別に算出したビジブルエリアの面積に各エリアの係数をかけ、合計する

（Fig.6）は、エリア別にカウントした面積とその比率を表す。各エリア別の面積の配分にはあまり差がないことが分かる。

（Fig.7）は各エリア別の係数をかけ、加えたものであるが、ビジブルエリアの差がより大きくなっていることが分かる。

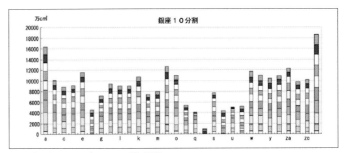

（Fig.6）分割された10個のエリア別の面積
係数をかける前の各エリア別面積の配分はほぼ同等である

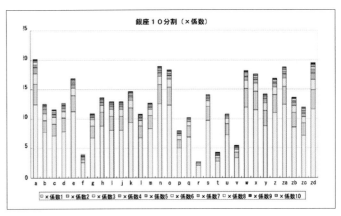

（Fig.7）各エリア別の係数をかけて加えた面積
街路の太さや観察地点まわりの空間の広さがビジブルエリアを大きく左右していることが分かる

街路空間のシークエンスを考えたとき、前後のつながりは重要である。面積をつないだときにできる折れ線、その傾き（変化率）こそが前後の関係を表し、空間の差を表すのに適した指標となると考えられる。この変化率はその大きさ（0からの距離）にその意味があり、±にはあまり意味がない。そのため変化率を絶対値で表す事でその大きさを比較する。

連続する2つの観測地点間において変化率は以下のように求める。

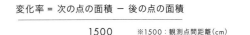

$$変化率 = \frac{次の点の面積 - 後の点の面積}{1500} \quad ※1500：観測点間距離(cm)$$

また、この変化率は、人の驚き具合などの人が街路空間において感じる心的変化と 何らかの相関性があると考えられる（Fig.8）。

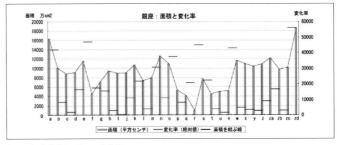

（Fig.8）ビジブルエリアの変化率
いわゆる微分係数であるが、これが大きいほど人間に与える心的変化が大きいことが予想される

対象街路における分析の実践と考察

　街路の形態の異なる3つの街路を選定し、解析法の実践を行った。対象地区は、1）岡山県高梁市、2）東京都世田谷区下北沢、3）東京都中央区銀座である（Fig.9）。

対象街路1　岡山県高梁市本町

ビジブルエリアの算出

考察
- ■ 結節点におけるビジブルエリアの面積の変化が大きい
- ■ a-b間、x-y間の変化率が大きく全体のまとまりを感じさせる
- ■ 空洞化した土地による連続的視野への影響が見られる

変化率と経路の関係

対象街路2　東京都世田谷区下北沢

ビジブルエリアの算出

考察
- ■ 街路が折れ曲がっているためビジブルエリアが135mの範囲まで達していない
- ■ 駅前広場による広がりがこの街路において象徴的である
- ■ v地点での道路幅員の変化によりまとまりが分断されている

変化率と経路の関係

対象街路3　東京都中央区銀座

ビジブルエリアの算出

考察
- ■ 地下鉄の出入り口によりビジブルエリアが制限されている
- ■ 昭和通りにより街路空間の流れが途切れる
- ■ 建物のセットバックによる面積が極端に少ない

変化率と経路の関係

（Fig.9）三街路におけるビジブルエリアの変化率の比較

心的変化との相関性

　分析によって導き出したビジブルエリア面積の変化率と人の心的変化との間に、何らかの相関性があることが予想されるため、下北沢の対象街路にて訪問者を対象者としたアンケート調査を行い、実際に人が感じる心的変化を抽出した。調査方法は自由記述式を採用し、対象者は地図を持ち、実際に街路空間を歩きながら心理的に変化を感じる点を地図にプロットし、その理由についても記述してもらった。また、対象者には2方向からのシークエンスについてアンケートに答えてもらった。解析により得られた変化率とアンケートにより得られた心的変化を比較することで、その相関性を示す。変化量のピークと心理的変化の場所はほぼ一致していた。「見える」範囲が急に変化することで、人間が少なからず心理的な変化を感じることを確認することができた（Fig.10）。

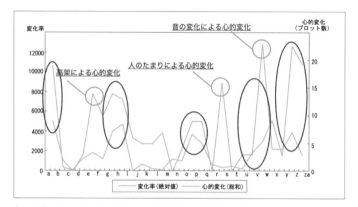

（Fig.10）下北沢地区のビジブルエリアの変化率と心的変化の相関性
ビジブルエリアが大きく変化している場所で
人が心的変化（驚きなど）を感じていることが分かる

Visualization of Streetscape Sequence by Light Source Projection Method

OUTLINE

In Chapter 1, we discussed the difference in the amount of information between unaware "visible" vision and conscious "seeing" vision. In this section, we deal with "visible" vision, which is the visual range by the light source projection method, and explore changes in the "visible" visual area by moving a human in a virtual space. We also simultaneously observe the amount of psychological change in response to this change, and explore the mechanism of the change that the visual sequence creates in the mind.

Concept of Visible Area

Vision is the most utilized information processing ability among humans. In most cases, cities are understood through "seeing," and people act by obtaining various information from a limited field of vision. Here, we define the "visible" or recognizable area as the visible area and analyze the visible area in the visual sequence of the street space by using the light source projection simulation method. Based on previous studies, the maximum distance at which a person's movement can be distinguished is set at 135 m, and a circle with a radius of 135 m from the point at which the person stands is analyzed as the area within which a person can recognize the street space (the visible area). Since there are various things in the street space that induce the act of "seeing," but they differ from person to person, it is thought that a method that only analyzes the visible area in a certain direction will result in a large difference from the state in which people are actually walking. Therefore, here, the visible area is analyzed as a 360-degree area centered on the person.

About the observation points

The path of movement in the street space is divided into fifteen-meter intervals. Observation points are established every fifteen meters and the visible area of each observation point is calculated. The observation points are the center of the street in streets without sidewalks and the center of the sidewalk in places with sidewalks, in order to pursue the visible area of people in the street space. This distance unit is the average distance an adult male walks in ten seconds, and since the movement is within the range of the near field of view, it is expected to allow for consideration at the human scale.

Visible Area and Visual Sequence

The change in the area of the Visible Area is calculated numerically using the light source projection method and analyzed with respect to the sequencing of the street space. Basically, the visible area is modeled by replacing people with point light sources and is shown as an area in which people can obtain spatial information visually. Then, changes in the visible area when a person moves are captured as a visual sequence.

Calculation of Visible Area by Light Source Projection Simulation Method

Visible Area Analysis

Visible area information obtained by the light source projection simulation method consists of 1) the form of the visible area and 2) the area of the visible area. The form of the visible area is pattern information with distance and size, and the area of the visible area is scalar information. In this section, the following analysis is performed for this information.

(1) Regarding the morphology of the visible area at the observation point, in addition to the "visible" landscape along the basically linear street, a large morphological change occurs when the vision opens up in a direction perpendicular to the direction of movement at T-junctions and intersections. The status of branching of the visible area, and the characteristics of the street can be seen in the length of the branches, especially on curved streets, where the size of the visible area does not reach 135 m, creating a landscape with a sense of enclosure.

(2) The area of the visible area and the amount of change in the visible area at each observation point were counted as the visible area multiplied by the attenuation coefficient corresponding to the distance, taking into account that the resolution of the human visual structure differs between nearby and distant objects and that it decreases in a gradational manner, similar to the attenuation of light. The visible area is counted by multiplying the attenuation coefficient according to the distance. First, the visible area is divided into 10 concentric circles, and areas 1 to 10 are designated from the inside.

Next, the ten divided visible areas are compared by applying different coefficients, taking into account the perspective from the observation point. Since the human eye can see by collecting optical information, we decided to use the coefficient of light attenuation =

1/square of the distance (attenuation of illuminance on the plane perpendicular to the light source). The distance from the observation point to the center of the area is defined as the distance and the coefficient is as follows.

Table in Fig. 5 assumes that the degree of visual attenuation is equivalent to the degree of light attenuation.

Fig. 6 presents the area counted for each area and its ratio. It can be seen that there is not much difference in the distribution of area by each area.

When considering the sequencing of street space, the relationship between the two consecutive points are important. The gap (rate of change) of the graph line formed when the areas are counted is considered to be a good indicator of the relationship between the two consecutive points and to show the difference in space. The rate of change has its meaning in its magnitude (distance from 0), and plus/minus has little meaning. Therefore, the magnitude is compared by expressing the rate of change as an absolute value.
The rate of change between two consecutive observation points is calculated as follows:
Rate of change = area of next point - area of next point /1500
*1500: distance between observation points (cm)

The rate of change is considered to have some correlation with the mental change that people feel in the street space, such as the degree of surprise of people (Fig. 8).

Practice and Discussion of the Analysis on the Target Streets

Three streets with different street forms were selected for investigation and analysis. The target districts were 1) Takahashi City, Okayama Prefecture, 2) Shimokitazawa, Setagaya-ku, Tokyo, and 3) Ginza, Chuo-ku, Tokyo.

Correlation with Psychological Change

Since it is expected that there is some correlation between the rate of change of the visible area and the psychological change of people, we conducted a questionnaire survey of visitors to the target streets in Shimokitazawa to extract the psychological change that people actually perceive. The survey method was an open-ended questionnaire, in which the subjects were asked to plot the points on the map where they felt psychological changes while actually walking along the street space, and to describe the reasons why they felt such changes. The subjects were also asked to answer a questionnaire about the sequence from two directions. By comparing the rate of change obtained by the analysis and the psychological change obtained by the questionnaire, we show the correlation between the two. The peak of the rate of change and the location of the psychological change were almost identical. We were able to confirm that a sudden change in the "visible" range causes a person to feel some amount of psychological change.

Fig. 1 Range of Visible Area (plane and elevation)
Fig. 2 Process for calculating the visible area
Fig. 3 Visible area at the observation point
Fig. 4 Calculation of the area of each area within the concentric circles
Fig. 5 Table assuming that the degree of visual attenuation is equivalent to the degree of light attenuation
Fig. 6 Area by 10 divided areas
Fig. 7 Area multiplied and added by the coefficient for each area
Fig. 8 Rate of change of visible area:
Fig. 9 Comparison of the rate of change of the visible area in the three streets
Fig.10 Correlation between the rate of change in the Visible Area and psychological change
        in the Shimokitazawa area

2. 建物の透過性と空間シークエンスを見える化する

概要

市街地における商業的街路は、住宅地の路地などと比較し、より公共性が高く、一般の来訪客が入りやすい空間となっている。特に都心部における商業的街路では、街路沿いには建物が途切れることなく配置され、建物ファサードが連続して並んでいる。「ウィンドウショッピング」とは、こうした商業的街路を歩きながら、沿道建物の連続するショーケースや店舗内を見て楽しむことを指すが、ここではそのメカニズムを考える。

沿道建物のファサードのデザインではガラスなどの透明な素材が多用され、ファサードの「透過性」が生じており、外部空間と内部空間の境界が曖昧になっている状況が見られる。このような「透過性」のあるファサードによる現象として注目すべき点は、「視線の透過による内部空間の外部化」であり、このことは、建築家・槇文彦（1928年～）が代官山のプロジェクト「ヒルサイドテラス」でも述べている「空間のヒダ」であるともいえよう（Fig.9）。ここでは、この曖昧な境界が商業的街路における歩行者の空間認識や心理変化に及ぼす影響に注目し、さまざまな街路において歩行者の移動により起こる「内部空間への視野の広がりによる街路空間のシークエンスの変化とリズム」に焦点を当てる。具体的にはこれを「知覚時間集積グラフ」により視覚化し、さらにアンケート調査により人々の心理変化との相関関係を明らかにする。

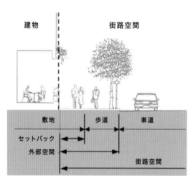

（Fig.1）街路と建物の透過性を示した概念図
道路の進行方向と直行する広がりにより
空間のヒダが生まれる

知覚時間量と、根拠としての歩行速度と可視領域

空間のシークエンスを体験すると、"ある風景が、ある時見え始めて、その後見えなくなる"という現象が途切れずに続いていることに気づく。それは、ある人間が移動中にその空間の連続の中で見たすべての風景が、「その人によって見られる時間」を有していることを意味している。言い換えれば、ある移動中の視点に対して、空間上のすべての地点は固有値である「知覚時間量」をもっているということである。

「知覚時間量」は、その人の歩行状態によって決まる。たとえば、歩行速度が速ければ、その歩行者が見る景色は速く流れ、知覚時間も短くなる。逆に歩行速度が遅ければ、知覚時間は長くなるだろう。しかし、街路空間には、目的をもって足早に歩き去る人、ゆっくりと散歩を楽しむ人、沿道の店を眺めながら街並みを楽しむ人などさまざまあり、歩

行速度は一定とは限らない。また、東京都市部の平均的歩行速度は1.56m/s（※国際交通安全学会による）という研究報告があるが、これは日本における主要な都市の中で2番目に速い速度である。（一位は大阪）このように都市によっても歩行速度が違うように、街路空間の性質によっても異なる。ここでは、足早に歩いている場合を除いて、散歩のような状態の歩行速度1.0m/sと設定する。

また、人は360度を一瞥することはできないが、注視による回頭行動などにより、人は常に同じ方向を見ていると限定することは難しい。そこで、「見える」視野角は360度とし、「可視領域」を"人が立つ地点から半径135mの円の内側"（先行研究より）とする。

知覚時間集積グラフの立ち上げとその検証

「知覚時間集積グラフ」とは、1人の歩行者が街路空間上を一定の速度で移動するとき、内部空間を含めた街路空間上の任意の点における「歩行者の視野に入る時間量」（知覚可能時間）の集積を示している。

まず、「内部空間への視野の広がり」とそれらが生み出すリズムに影響を与えている空間構成要素として、（1）建物の間口（開口）幅、（2）道路の幅員、（3）街路の形状、および（4）ファサードの透過性、の4つを抽出し、これらの要素について、実際に複数の商業街路で調査を行った（Fig.2, 3）。このうち（4）ファサードの透過性に関しては、レベルに応じた5段階の重み付け評価を行っている。ここでは、東京・下北沢南口商店街通り、銀座中央通（5～6丁目）における調査の事例を示す。

五段階評価	Level 1	Level 2	Level 3	Level 4	Level 5
下北沢					
係数	1.0	0.75	0.5	0.25	0.0

（Fig.2）建物ファサードの透過性を5段階に評価し係数を与える
level1は開放的な店舗、level5は閉鎖的な店舗とし
その間を透過性の度合いによって分類し、係数を与えた

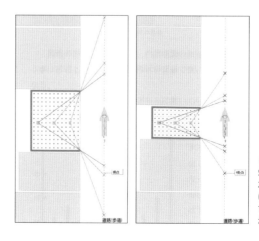

（Fig.3）間口幅による知覚時間の違い
左は間口7,000mm、右は3,000mmの場合を示す一般に間口が広い方が知覚時間は長くなる

GISの3D機能による「知覚時間集積グラフ」の作成

以下に、知覚時間集積グラフを立上げる具体的な方法と、下北沢地区、銀座地区の事例を示す（Fig.4, 5, 6）。

1	二次元地図の作成

人の目線の高さにおいて分析するため、1,500mmの高さにおける建物形状を、10cm単位でで実測する。

2	視点の設定

歩行者がこの軸上を移動すると仮定して、道路（歩道）の中心軸上に視点を設置する。この場合、実際は視点が無数に存在するのだが、今回は5,000mm毎に視点を設置した。
この距離が短くなればなるほど、より実際の歩行状況に近いといえる。

視点

3	視点の設定

作成した二次元地図に建物や塀など双線を遮るものの高さの情報を与え、三次元地図に変換する。光を投射する際に透過しないように考慮する。この時ファサードは高さを与えない。

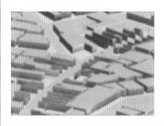

4	光の投射

視点の位置に点光源を設置し、光を投射する。光が当たっている部分が視野範囲となる。
街路空間における視覚認識は輝度の差異によって表される。

5	可視領域の作図

4で作成した画像をもとに可視領域の図形を作成する。半径135mに含まれる部分が可視領域となる。

可視領域

6	ファサードでの分割

5で得た可視領域の図形を建築ファサードで分割する。ファサードから内側に入った内部領域部分にはその透過度に従った係数がかけられる。

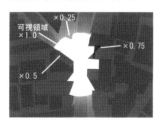
可視領域
×0.25
×1.0
×0.75
×0.5

7	地図上のグリッドに情報を入力

GISソフトを用いて、測定エリアを500mmグリッドに分割する（グリッド一個を街路空間上の任意の点と位置付ける）。各視点毎に、可視領域に含まれるグリッドを選定し、グリッドに「可視領域か否か」についての情報を入力する。

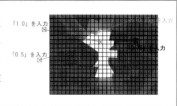
「1.0」を入力
「0.5」を入力

（Fig.4）「知覚時間集積グラフ」を立ち上げるプロセス

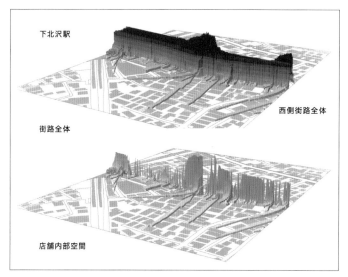
下北沢駅
街路全体
西側街路全体
店舗内部空間

（Fig.5）下北沢地区の「知覚時間集積グラフ」
上の図は道路を含んでいるため、高いグラフ面になっているが、中間の四差路あたりの集積度が高いことが分かる。下の図は、建物内部だけのグラフだが、駅に近い沿道建物の透過性が少なく、遠く行くに連れて多いことが分かる。

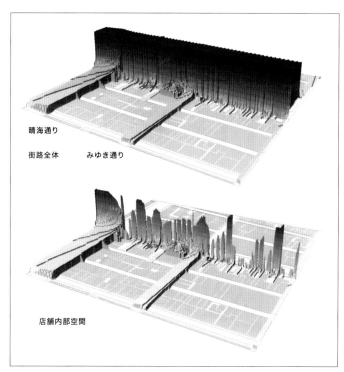
晴海通り
街路全体　　みゆき通り
店舗内部空間

（Fig.6）銀座地区の「知覚時間集積グラフ」
上の図では、一様に高いグラフ面となっており、視野変化が少ないことを示している。左端の太いグラフは、交差する晴海通り方向の視界である。下の図は、晴海通りからみゆき通りにかけて、透過性の高い店舗が並んでおり、みゆき通りより西側のエリアの方が若干閉鎖的であることが分かる。

「知覚時間集積グラフ」と体験する人の心理の関係

アンケートを実施し、空間が歩行者の心理に何らかの変化を与えるポイントを「心理変化点」とし抽出した。その心理変化点と、知覚時間集積グラフでの分析によって得られた奥行きを感じる知覚時間量の差異との関係性を検証する。

（Fig.7）下北沢の測定街路を除いた知覚時間集積グラフ
色が濃くなっている部分が歩行者の視線が集よりやすい箇所を示している

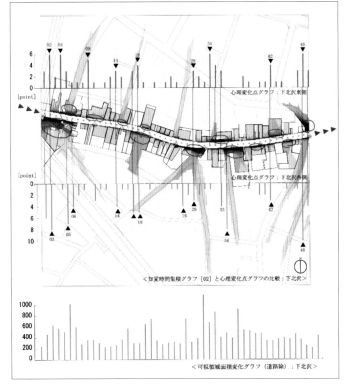

（Fig.8）下北沢の知覚時間集積グラフと心理変化グラフの比較
実際の心理変化点と知覚集積ポイントの高い箇所がほぼ一致していることが分かる

　下北沢の分析図において、奥行きを感じている時間の長い沿道の内部空間をピックアップした（赤丸内）が、視野が急激に広がる結節点が多く含まれていることが分かる（Fig.8）。これは、その前後において奥行きに対する知覚時間量の差異が生まれていることを意味している。心理変化点と比べてみると、ある程度グラフが一致しており、そこで心理変化が起こっていることが明確となった。ここで注目すべき点は、奥行きを感じる時間の長いポイントの中に結節点が多く含まれていることである。

　「知覚時間集積グラフ」により、移動中の歩行者の視野の集積度の分布を、わかりやすく視覚化することができた。以上の結果を踏まえて、視野の集積度の分布、内部空間の奥行きを感じる時間などの視野変化について、設計段階から把握し、計画していくことの意味は大きい。また、「知覚時間集積グラフ」とアンケートによって抽出した心理変化点との比較考察により、内部空間の奥行きを感じる時間の変化と差異が、心理に何らかの影饗を与えている要因の1つであることが確認された。

Visualization of the Building Transparency and Spatial Sequence

OUTLINE

Commercial streets in urban areas are more public and easily accessible to general visitors compared to alleys in residential areas. In particular, commercial streets in urban areas have an uninterrupted arrangement of buildings along the street and a continuous line of building facades.

Transparent materials such as glass are used extensively in the design of roadside building facades, creating a "permeability" of the facade and blurring the boundary between the exterior and interior spaces. One of the phenomena that should be noted with such "permeable" facades is the "externalization of interior space through the transmission of the line of sight," which can be described as the "pleats of space" described by architect Fumihiko Maki (1928-) in his "Hillside Terrace" project in Daikanyama, Tokyo. Here, we focus on the effects of these ambiguous boundaries on pedestrians' spatial perceptions and psychological changes in commercial streets, and the "changes and rhythms in the sequence of street spaces caused by the expansion of the field of vision into the interior space" that occur as pedestrians move along various streets. Specifically, this will be visualized using a "perceptual time accumulation graph," and a questionnaire survey will be conducted to clarify the correlation between this graph and people's psychological changes.

Perceived Time Volume, Walking Speed and Visible Area as the Basis

When we experience the sequence of space, we notice that the phenomenon of "a certain landscape becoming visible at a certain time and then becoming invisible" continues uninterruptedly. This means that every point in space during their movement has an eigenvalue, a "perceived amount of time," with respect to a certain moving point of view.

The "perceived amount of time" is determined by a person's walking condition. For example, if a pedestrian walks at a high speed, the scenery they see flows quickly and the perception time is shortened. Conversely, if the walking speed is slow, the perceived time will be long. However, walking speed is not always constant and the average speed in urban areas differs from city to city, which are also affected by the nature of the street space. Here, we set the walking speed as 1.0 m/s in a stroll-like condition.

In addition, although a person cannot take a 360-degree glance, it is difficult to limit that a person always looks in the same direction due to turning behavior caused by gazing. Therefore, the "visible" viewing angle is set to 360 degrees, and the "visible area" is set to "inside a circle with a radius of 135 m from the point where the person stands" (from advance research).

Launch and Verification of Perceived Time Accumulation Graphs

The "perceptual time accumulation graph" indicates the accumulation of "the amount of time in the pedestrian's field of vision" (perceptual time) at any point on the street space, including the interior space.

First, four spatial components were extracted as influencing the "expansion of the field of vision into the interior space" and the rhythm they create: (1) the width of the building frontage (opening), (2) the width of the street, (3) the shape of the street, and (4) the permeability of the facade, and the study of these components were conducted on several commercial streets. Of these, for (4) facade transparency, a five-level weighted evaluation is used according to the level of transparency. Here, we show an example of the survey conducted on Shimokitazawa South Exit Shopping Street and Ginza Chuo-dori (5-6 chome) in Tokyo.

Creation of "Perceived Time Accumulation Graph" using the 3D function of GIS

Below is a concrete method for creating a perceptual time accumulation graph and examples from the Shimokitazawa and Ginza districts.

The graph surface is uniformly high in the upper figure, indicating little change in the field of view. The thick graph on the far left is the field of view in the direction of the intersecting Harumi-dori Avenue. The lower figure shows that the area from Harumi-dori to Miyuki-dori is lined with highly transparent stores, and the area west of Miyuki-dori is slightly more closed than Miyuki-dori.

Relationship between the "perceived time accumulation graph" and the psychology of people experiencing the space

We conducted a questionnaire survey and extracted "psychological change points," which are the points where the space influence shifts in the psychology of pedestrians, and verified its relationship with the difference in the amount of time spent perceiving depth.

In the analysis chart for Shimokitazawa, we picked the interior space along the road where the perception of depth is long (inside the red circle). We can see that it contains many nodal points where the field of vision rapidly expands. This means that the difference in the amount of time spent perceiving depth was generated before and after the nodal point. When compared to the psychological change points, the graphs are consistent to some extent, and it is clear that psychological changes are occurring there. It is noteworthy here that many nodal points are included in the points where the perception of depth is long.

The "Perceived Time Accumulation Graph" enabled us to visualize the distribution of the degree of accumulation of the moving pedestrian's field of vision in an easy manner. Based on the above results, it is highly significant to understand and plan for changes in the field of vision, such as the distribution of the degree of integration of the field of vision and the time it takes to perceive the depth of the interior space, from the design stage. In addition, through comparison of the "perception time accumulation graph" and the psychological change points extracted by the questionnaire, it was confirmed that the change and difference in the time to perceive the depth of the interior space is one of the factors that have some influence on the psychology..

Fig. 1　Conceptual diagram showing the permeability of streets and buildings
The expanse perpendicular to the direction of the street creates pleats in the space
Fig. 2　Five levels of permeability of building facades are evaluated and given coefficients
Level 1 is for open stores and level 5 is for closed stores, and the coefficients are assigned according to the degree of permeability between them
Fig. 3　Difference in perceived time according to frontage width
The left panel shows a frontage of 7,000 mm and the right panel shows a frontage of 3,000 mm. In general, the wider the frontage, the longer the perceived time
Fig. 4　Process of creating a "perceptual time accumulation graph
Fig. 5　"Perceptual time accumulation graph" in Shimokitazawa district
Fig. 6　"Perceived Time Accumulation Graph" for the Ginza area
Fig. 7　Perceived time accumulation graph excluding the measured streets in Shimokitazawa
Fig. 8　Comparison of the perceptual time accumulation graph and the psychological change graph in Shimokitazawa

1. 建物の内部空間の視覚シークエンスを見える化する

概要

前頁で光源投射法による視覚範囲の分析が主として「見える」視覚を扱っていたのに対し、ここでは、点群（3次元座標を持った点の集合で3D測量などで得られる情報）により、実空間をデジタル化したデータに変換し、その仮想空間の中で人間を移動させ、ある方向を「見る」視覚領域の変化を探る。また、この変化に対する心理的変化量を同時に観察し、視覚的シークエンスが心理に与える変化のメカニズムを探る。

点群による空間の記述

点群データはレーザー光線の往復によって得られる実空間内の位置座標を示すが、今まで記述が難しかった物質の物理的な位置関係がすべてデジタルデータで扱えるようになったことが画期的な進化である。しかし、写真によるフォトグラメトリーや3Dスキャナによる点群データは膨大な数の点から成り立っており（通常の部屋レベルで数十万、建物全体や外部空間では億単位の点の数）、各点それぞれにx,y,zスキャナ情報が入力されているためデータ量は莫大になりがちである。RhinocerosやGrasshopperを用いてひとつの点データに関して処理行うためには、データをある程度削ぎ落として圧縮することが重要となる。通常は、15~30cm角くらいの立方体の集積に変換して、データの処理を行うことが多い。これをVoxel化するという（Fig.2）。

次に、空間分析を現実に近づけるために、モデル上に人を配置し、誘導視野内の点群について情報処理を行なう。

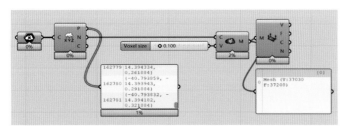

(Fig.1) Grasshopperによるプログラミング

(Fig.2) Voxelデータによる中野キャンパスアトリウムのイメージ

点群による「見る」室内空間のシークエンス

ここでは、視覚的なシークエンスに沿った視野角の変化（空間の体積量の変化）と開放感の感覚量の関係性を探る。第一段階として、人間の視野角形態としてラッパ型を想定し、その変化をどのように表現するかを検討する（Fig.3）。次に、第二段階として、その変化と心理的変化の相関関係を探る。そして、第三段階としては、視覚的奥行き感のあり方を室内外で検討し、建築家・槇文彦（1928年～）による「奥性」の概念を具体的に探る。第一段階については、明治大学中野キャンパス、第二段階については、国際文化会館（港区六本木）、第三段階については、代官山ヒルサイドテラスを対象に分析する。これらの対象建物は、公共性が高く、かつ変化に富んだ空間構成を示しているため選定した。

(Fig.3) 人が「見る」ことを
シミュレーションしたラッパ型の視野範囲
このラッパ型形態の変化をシークエンスと位置づけ、分析を進める

1) 中野キャンパスアトリウムの視覚シークエンス分析

ここでは、明治大学中野キャンパスのエントランス周りに移動経路を設定し、それに沿って、視覚の変化を記述し、分析を行った。

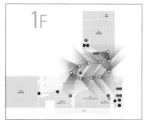

(Fig.4) 3Dスキャナによるエントランスホールと移動経路

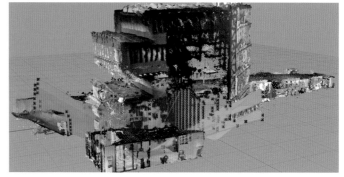

(Fig.5) 3Dスキャナによる明治大学中野キャンパスの外観

(Fig.6) 移動した向きに沿ったシークエンス

(Fig.7) シークエンスを直線に変換したイメージ ラッパ状の形態が、移動する向きによって、刻々と変化していることが分かる

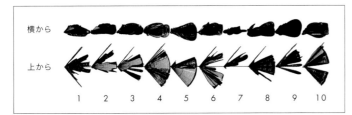

(Fig.8) 移動距離と視野範囲の変化を示した図
ここでの視野角は、誘導視野範囲である中心から左右に50度の範囲を用い、その距離は30mとした

人を取り巻く、壁や天井など制約により、視野範囲が変化し、豊かなシークエンスが生まれていることが分かった（Fig.6,7,8）。

2) 国際文化会館における
視覚シークエンス分析と心理的変化

国際文化会館（Photo1, 2）の内部を移動し、3Dスキャナを利用して点群データを取得し、Voxelデータ化する。次に、移動経路に沿ったラッパ型の体積の変化をグラフ化する（Fig.9,10,11,12）。その上、「開

(Photo1) 国際文化会館

(Fig.10) 移動経路

(Photo2) 国際文化会館内部の写真

(Fig.9) 国際文化会館の外観のデジタルデータ

放感」の感覚量を「開放感に関するアンケート」を実施する形式で評価する（Fig.12）。アンケートの被験者は建物を訪れたことのない人とし、建築内部に設定した歩行ルートを身長165cmの人間の目線の高さから録画した動画を視聴する方法でアンケートを実施した。

アンケートの内容は、歩行ルートに等間隔に30ポイント設けて、ポイントごとに「開放感」を1〜5段階で評価する。

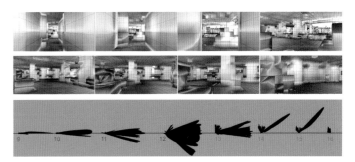

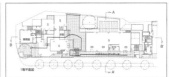

(Fig.11) Voxelイメージと上から見たラッパ型の変化

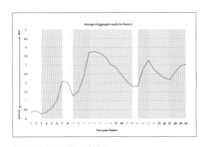

(Fig.12) ラッパ形の変化と心理的変化を重ねたグラフ
廊下から広いラウンジに出た11→12のシーンで大きく変化が起きていることが分かる

シークエンスの変化と心理的変化をグラフ上で照合することで、視野範囲が大きく変化する箇所では大きな心理的変化が起きていることを裏付けできた。しかし、これはビデオという共通の視覚変化で実験した結果なので、実際には、頭が動く環境での検証も重要である。

3) 代官山ヒルサイドテラスにおける
視覚シークエンス分析と奥性

建築家槇文彦の初期の作品であり、彼の「奥性」の思想が最も良く示されている代官山ヒルサイドテラスは、1期に始まり7期に至るまで、約40年間をかけて開発された複合施設である。ここで、A・B棟およびC棟を対象に、室内から室外に至る経路を設定し、視覚領域を示す

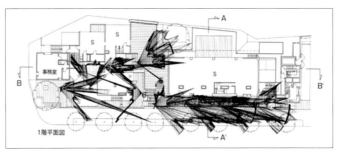

(Fig.13) 代官山ヒルサイドテラス アクソメ図　　(Fig.14) 測定した経路（A・B棟）

(Fig.15) 視覚領域のラッパ型の変形過程（A・B棟）

ラッパ型がどのように変化し、どこで最も「奥性」を感じる見通しの良い視覚が得られるのかを、実地調査を通して確認した（Fig.13,14,15, 16,17）。

このヒルサイドテラスは通常の箱型の商業施設ではなく、開放と閉鎖が繰り返され、微地形といわれる微妙なレベル差が設定されているため、人が動く時には、常に足元と前方に視線を動かしながら、移動するように自然と促される。また、それらの視覚の変化は、強制的ではなく、適度な刺激を与えてくれるので、ある種のわくわく感にも通じる心理的変化を得ることができる。

CHAPTER 2, 4-1 SPATIAL DESCRIPTION AND SEQUENTIAL ANALYSIS USING POINT CLOUDS

Visualization of the Visual Sequence of a Building's Interior Space

SUMMARY

While the analysis of visual range by the light source projection method dealt mainly with "visible" vision, here, a real space is first converted into digitalized data using a point cloud (explained in detail below), then a human being moves in the virtual space to explore changes in the visual area of "seeing" in a certain direction. We will also simultaneously observe the amount of psychological influences in response to this change, and explore the mechanism of the visual sequence that gives to the mind.

Point Cloud Data

Point cloud data represent positional coordinates in real space obtained by reciprocating laser beams, which enables all physical positional relationships of materials handled as digital data. However, photogrammetric and 3D scanner point cloud data consists of an enormous number of points (hundreds of thousands of points at the level of an ordinary room, and hundreds of millions of points for an entire building or external space). To process a single point data using Rhinoceros or Grasshopper, it is important to reduce and compress the data, which are converted into an accumulation of 15 to 30 cm square cubes. This is called Voxelization.

In order to bring the spatial analysis closer to reality, a person is placed on the model and information processing is performed on the point clouds in the guided field of view.

Sequence of indoor space "seen" using the point cloud

In this section, we will explore the relationship between changes in the viewing angles or volumes of space along the visual sequence, and the sensory amount of openness. As the first step, we assume a trumpet shape as form of human viewing angle and examine the changes in expression. As the second step, we will explore the correlation between the changes of human viewing and psychology. As the third step, we will examine the visual sense of depth for both indoors and outdoors, and specifically explore the concept of "depth" by architect Fumihiko Maki (1928-). Different types of highly public buildings with varied spatial composition were selected for examination: Nakano Campus of Meiji University for the first stage, International House of Japan (Roppongi, Minato-ku) for the second stage, and Daikanyama Hillside Terrace for the third stage.

(1) Sequence analysis of the Nakano Campus Atrium (description of visual sequence)

In this section, a movement path was set up around the entrance of the Nakano Campus of Meiji University, and visual changes were described and analyzed along the path.

It was found that constraints such as walls and ceilings surrounding people changed the range of vision, creating a rich sequence.

(2) Sequence analysis in the International House of Japan (visual sequence and psychological change)

The participants move around the interior of the International House of Japan using a 3D scanner, and the data is converted into Voxel data. Next, changes in the trumpet-shaped volume along the path of movement are graphed. The sensory volume of "openness" is then evaluated in the form of a "questionnaire on openness." The questionnaire respondents were targeted for those who had never visited the building, and were requested to watch a video recorded from the eye level of a person 165 cm tall on a walking route set up inside the building.

The questionnaire consisted of 30 points equally spaced along the walking route, with each point rated on a scale of 1 to 5 for "openness.

By matching the graphs showing changes of sequence and psychology, we were able to corroborate large psychological changes occurring at points at which the range of vision shifts significantly. However, since this is the result of an experiment using video, which has restricted visual change, it is important to verify the results in an environment where people's heads can easily move.

(3) Sequence analysis at Daikanyama Hillside Terrace (depth visualization)

Daikanyama Hillside Terrace, an early work of architect Fumihiko Maki and the best example of his "depth" philosophy, is a complex facility developed over a period of about 40 years in various phases. Here, we set up a route from the interior to the exterior for Buildings A, B, and C and confirmed how the trumpet shape would change, and where a sense of "depth" can be obtained.

Considerations

This Hillside Terrace is not an ordinary box-shaped commercial facility and has an arrangement of openness and closedness with subtle level differences, known as microtopography, which encourages people to naturally traverse around the space making them always move with their eyes to their feet and ahead. These visual changes are not forced, but rather provide a moderate stimulation, which can lead to a psychological change that can be likened to a certain sense of excitement.

Fig.1 Programming with Grasshopper
Fig.2 Image of the Meiji University Nakano Campus Atrium using Voxel data
Fig.3 Viewing area in the shape of a trumpet that simulates what people see
Fig.4 Entrance hall and movement path by 3D scanner
Fig.5 Exterior view of Meiji University Nakano Campus using 3D scanner
Fig.6 Sequence along the moving direction
Fig.7 Sequence converted to a straight line
Fig.8 Variation of moving distance and field of view
Fig.9 Digital data of the exterior of the International House of Japan
Fig.10 Moving route
Photo1 Exterior view of the International House of Japan
Photo2: Inside of the International House of Japan
Fig.11 Voxel image and change of trumpet shape seen from above
Fig.12 Overlaid graph of the change in trumpet shape and psychological change

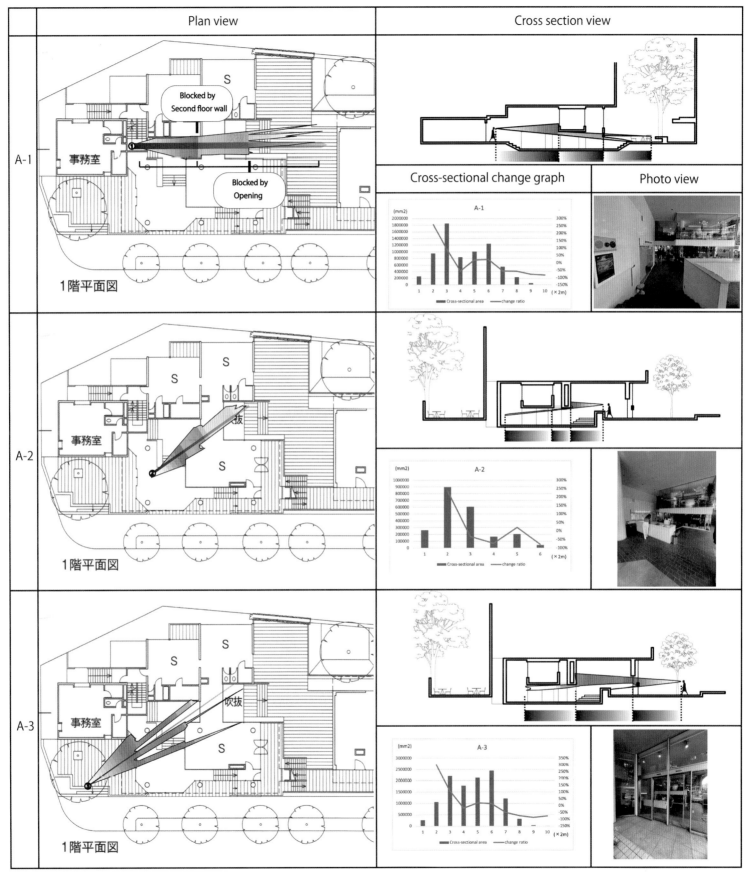

（Fig.16）「奥性」を示すラッパ型の変形過程（A・B棟）

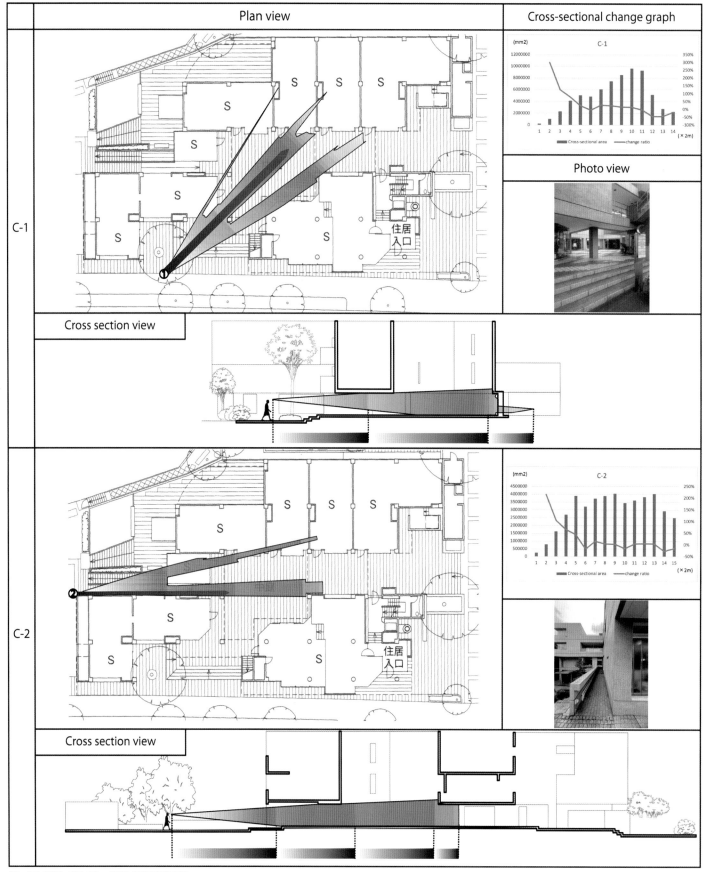

	Plan view	Cross-sectional change graph
C-1	S S S S S S S S 住居 入口	C-1
	Cross section view	Photo view
C-2	S S S S S S S 住居 入口	C-2
	Cross section view	

(Fig.17)「奥性」を示すラッパ型の変形過程（C棟）

2. 街路空間の視覚シークエンスを見える化する

概要

わが国には、路地や商店街のような特徴のある線状（リニアー）の街路が存在し、多くの人々を惹き付けてきた。しかし、近年展開されている都市再開発事業では敷地が巨大街区（スーパーブロック）となりがちで、街路で重要といわれる豊かなシークエンスのある空間体験はあまり実現されていない。ここでは、歴史的財産である花街建築や路地裏を有する東京・神楽坂地区と、複雑な街路構成と商店街を有する東京・下北沢地区を対象に、視野体積変化のシークエンスから街路の特徴を掴む。

観察の基本的ルール

1) 視野角については、関係資料を参考に、水平方向に100度、垂直方向に上下それぞれ50度、35度と設定した。また、視野距離については、道路幅員が2〜3mである神楽坂では視野距離を10mと設定し、道路幅員が5〜6mである下北沢では30mと設定した。

（Fig.1）ラッパ形状の「見る」視野範囲
この形状の変化を調べ、移動中の空間シークエンスの変化を探る

2) 歩行路の往路・復路を別々に分析し、往路と復路に違いが生じるかを調べるため、街路空間に歩行路を設定し、ルートの長さと視野距離を考慮し、神楽坂は5mごと、下北沢では10mごとに観測視点を設定した。

1) 神楽坂のシークエンスを見える化する

神楽坂の視野体積については、道幅が狭く、建築のセットバックや植栽の影響を強く受けていたため、数値が細かく上下しており、ルートの中間地点に向かって視野体積が低くなっていた。植栽の形状・塀が雁行していることから、往路と復路で計測した視野体積に違いも生じている。

（Photo1）神楽坂の路地景観

（Fig.2）神楽坂の路地裏における
調査対象地（平面）

（Fig.3）神楽坂の路地裏における測定地

（Fig.4）神楽坂の連続的ラッパ形状

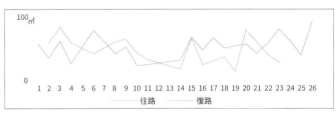

（Fig.5）神楽坂の「見る」視野範囲の変化。往路と復路で若干の違いがみられる

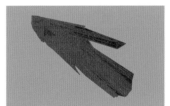

（Fig.6）ラッパ形状の変動要因
電柱や看板などで視野がけられることが多い

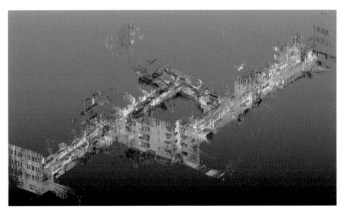

（Fig.7）神楽坂地区の街路のデジタルデータ

2）下北沢のシークエンスを見える化する

視野体積は、地点0-1、7、20-21で高い数値が読み取れ、地点11-15では連続的なファサードにより視野体積の変化が少ない。また、地形の起伏の影響で下り坂である往路の方が復路より数値が高い結果が得られた。

（Photo2）下北沢の商店街の景観

（Fig.8）下北沢南口商店街における
調査対象地（平面）

（Fig.9）下北沢南口商店街における測定地

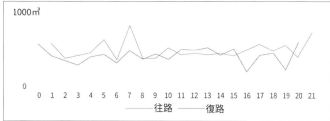

（Fig.10）下北沢の「見る」視野範囲の変化
一部、T字路で、往路と復路で大きく違う場所が見られる

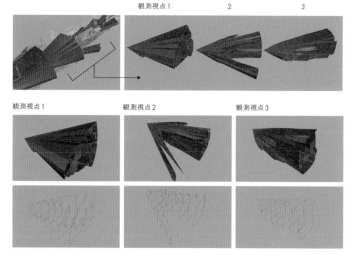

観測視点1　　　　2　　　　3

観測視点1　　　　観測視点2　　　　観測視点3

（Fig.12）下北沢の観測視点1〜3のラッパ形状
T字路の部分でラッパ形状が乱れていることが分かる

街路と建物の内部空間の視覚シークエンスの大きな違いは、街路内では基本的に上下移動が少なく空も視野に入るため、視線は自然と上方向に向かう傾向が強く、逆に国際文化会館（61頁）やヒルサイドテラス（61〜64頁）などの室内区間では、天井下で上下移動もあるため、視線は下方気味になる傾向が強い点である。外部空間であっても、上下移動が大きい階段や坂道（例えば、ローマのスペイン階段など）は景観的な変化が大きく印象に残りやすいため、映画や小説の舞台になりやすい。

神楽坂も下北沢も路地（細い道）で有名であるが、神楽坂は道路線形が直線的で、料亭のまちといわれるように私的性格が強く、看板や店舗からのにじみ出しが少ない。一方、下北沢は道路線形が湾曲しており、逆に公共性が高い商業地あるため、看板や店舗の付属物が多い。当初、下北沢の方がラッパ型のボリュームの変形が大きいかと予想されたが、結果としては、神楽坂のほうが塀や奥の建物の見え方が大きく変化し、下北沢の方が店舗が一様に並んでいるのでラッパ型のボリュームの変形は小さかった。槇文彦の概念で言えば、神楽坂のほうが「空間のヒダ」が多いと言える。

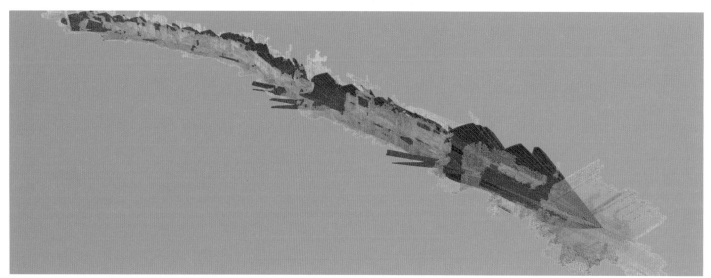

（Fig.11）下北沢の連続的ラッパ形状

Visualization of the Visual Sequence of the Street Space

SUMMARY

In Japan, linear streets with distinctive characteristics, such as alleys and shopping streets, attract people. However, recent urban redevelopment projects tend to create huge blocks of land, and the richly sequenced spatial experience that is considered to be important on the street is often not realized well. Here, we will take the Kagurazaka and the Shimokitazawa district of Tokyo in order to identify the characteristics from the sequence of changes in the volume of vision: the former with its historical assets of hanamachi architecture and back alleys, and the latter with its complex street configuration and shopping streets.

Basic rules of observation

(1) For the viewing angle, we set 100 degrees horizontally, and 50 and 35 degrees for vertical upper and lower way respectively, in reference to advance data. Then, the viewing distance was set at 10 m in Kagurazaka, where the street width is 2 to 3 m, and 30 m in Shimokitazawa, where the street width is 5 to 6 m.

(2) In order to separately analyze outbound and inbound walking routes and to examine whether differences occur between the two, walking routes were set along the street space with observation viewpoints at every 5 m in Kagurazaka and every 10 m in Shimokitazawa, considering route length and visual field distance.

Examine changes in this shape and explore the spatial sequence during movement

(1) Visualizing the sequence of Kagurazaka

Regarding the visual field volume of Kagurazaka, the values went up and down due to the narrow width of the street and the strong influence of architectural setbacks and plantings, and became lower toward the midpoint of the route. The differences between both routes were also affected by the shape of plantings and the zigzag pattern fences.

(2) Visualizing the Shimokitazawa Sequence

High values for visual field volume can be read at points 0-1, 7, and 20-21, and little variation at points 11-15 due to the continuous facade. In addition, higher values were obtained on the outward slope, which is downhill, than on the return slope due to the undulating topography.

The major difference between the visual sequence in the street and in the building interior, is that, there is little vertical movement in the street and one naturally inclines to look upward because the sky is within their field of view. The "up" and "down" directions of the viewer's eye are also strong in exterior spaces with stairways and slopes that has large vertical movements (e.g., the Spanish Steps in Rome). Such places can be popular as movie settings and novel sceneries because of large changes in the townscape that gives strong impression to people.

Both Kagurazaka and Shimokitazawa are famous for their alleys, but Kagurazaka has a straight road alignment and a strong personalized character, as it is known to be a ryotei (traditional Japanese restaurant) town, with few signs or encroachments from the stores. In contrast, Shimokitazawa has curved streets and is a highly public commercial area with many signboards and store fixtures. Initially, Shimokitazawa was expected to have a larger deformation of the trumpet-shaped volume, however, it turned out to result with Kagurazaka, which was influenced by significant changes caused by the appearance of fences and the buildings behind. In this sense, stores in Shimokitazawa are uniformly lined up. According to Fumihiko Maki's concept, Kagurazaka has more "folds of space."

Fig. 1　"Viewing" field of view range of the trumpet shape
Fig.2　Surveyed area in the back alley of Kagurazaka (plan)
Fig.3　Measured area in the back alley of Kagurazaka
Fig.4　Continuous trumpet shape in Kagurazaka
Fig.5　Change of "viewing" view range of Kagurazaka
Photo1　Alley townscape of Kagurazaka
Fig.6　Cross section of trumpet shape
Fig.7　Digital data of the street at Kagurazaka district
Photo2　View of the shopping street in Shimokitazawa
Fig.8　Surveyed area in the Shimokitazawa south exit shopping street (plan)
Fig.9　Measured location in the Shimokitazawa south exit shopping street
Fig.10　Change of "viewing" view range in Shimokitazawa
Fig.11　Continuous trumpet shape in Shimokitazawa

3. 建物の内部空間の色彩を見える化する

概要

地域や建物の環境情報として、色彩は重要な要素の一つである。自然環境の色彩が私たちに与える環境も重要であるが、建物内部の色彩も私たちの日常の活動を包む空間であるので、同様に重要である。ここでは、いくつかの建物の内部空間について色彩分布の分析を行い、空間の特性について色彩を軸に追及している。

点群による建物内部の色彩分布の確認と分析

ここでは、室内の色彩について、点群を用いて、色彩の分布度をシミュレーションする。
1）まず、点群データを30cmのVoxelデータに変換する。
2）点群データの色情報はRGB形式で表示されるため、HSV形式に変換して画像にする（Fig1,2,3）。

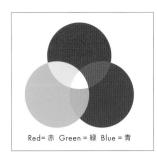

（Fig.1）RGB形式の構成

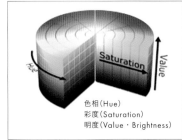

色相（Hue）
彩度（Saturation）
明度（Value・Brightness）

（Fig.2）HSV形式の構成

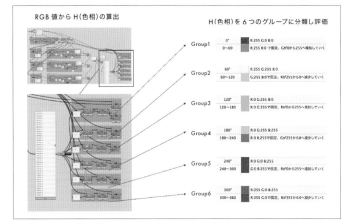

（Fig.3）グラスホッパーにより色相を分解するプログラム

明治大学の中野キャンパスのアトリウムと生田キャンパスの都市建築デザイン研究室を対象に分析を行った。色相の図を見ると、大学キャンパスのアトリウムも研究室も多様な数値の色相からなっている

ことがわかる。アトリウムは色相分布が集中的である（Fig4,5,6）。一方、研究室は離散的である（Fig7,8,9）。また、彩度を見ると、アトリウムは集中的な彩度を持っているが、研究室は彩度分布も離散的である。

測定を行った2つの地点はそれぞれ、アトリウムはパブリック性が強く、研究室はプライベート性が強い空間であると位置づけることができる。色相、彩度の円グラフとその分布から考察を行うと、パブリック性が強い空間ほど高彩度で、ある特定の色相範囲に点が偏って密集しており、プライベートな空間ほど、多様な数値の色相が空間内に離散的に存在している傾向があることが分かった。

（Fig.4）明治大学中野キャンパスアトリウムの30cmのVoxel図

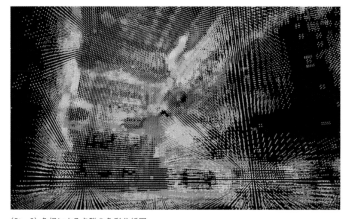

（Fig.5）色相による点群の色彩分析図
全体に高彩度の色が使われ、特にアクセントカラーといわれる赤色が中央のエスカレーターに象徴的に使用されている

（Fig.6）明度による点群の色彩分析図

（Fig.7）明治大学都市建築デザイン研究室の30cmのVoxel図
基本的には、トップライト、カーテンウォールなどの部位の明度が高いことは分かるが
中央のエスカレーターの側面も明度に貢献していることが分かる

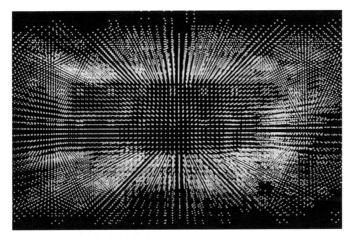

（Fig.8）色相による点群の色彩分析図
床面に暖色系が使われ、壁や天井部は寒色系に色彩が分布している

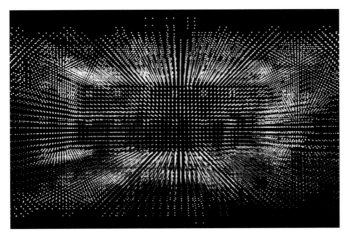

（Fig.9）明度による点群の色彩分析図
明度的には、天井・壁部が明るく、中央のテーブルの上も明るくなっていることが分かる

　点群による3Dスキャンの問題点の一つとして、ソリッドな壁ではない部分の透明ガラスの面やカーテンウォールなどの部分でレーザー光が飛び出してしまうので、うまく奥行き感などを表現できない、データが重いので簡単にその場で処理できない、などの点があげられる。しかし、内外部の空間のデジタル化という意味では、飛躍的な技術的進歩を遂げたことは間違いない。

Visualizing of the Colors of a Building's Interior Spaces

SUMMARY

Color is one of the most important elements of environmental information about an area or building. While colors of the natural environment are important contextual information, colors inside buildings are equally important because they are the spaces that surround our daily activities. Here, we analyze the distribution of colors in interior spaces of several buildings, pursuing the characteristics of space based on color.

Confirmation and analysis of color distribution
inside buildings using point clouds

In this section, we use point clouds to simulate the degree of color distribution in interior spaces.
First, the point cloud data is converted to 30 cm Voxel data.
Since the color information of the point cloud data is displayed in RGB format, it is converted into an image in HSV format (Fig. 3).

The analysis was conducted on the atrium of the Nakano Campus and the Urban Architectural Design Laboratory of the Ikuta Campus of Meiji University. The hue diagram shows that both the atrium and the laboratory on the university campus consist of a variety of numerical hues. The hue distribution in the atrium is intensive, while that in the laboratories is discrete. In terms of saturation, the atrium has an intensive saturation, while the laboratory has a discrete saturation distribution.

The atrium is a more public charactered space and the laboratory is a more private charactered space. The hue and saturation pie charts and their distributions indicate that the more public the space is, the higher the saturation and the more points are concentrated in a certain hue range, and that the more private the space is, the more discrete the hues of various values tend to be within the space.

Basically, we can see that the top light, curtain wall, and other parts of the building have high brightness, but we can also see that the sides of the central escalator also contribute to the brightness of the building.

Warm colors are used on the floor surfaces, and cold colors are distributed on the walls and ceiling areas.

The ceiling and walls are brighter, and the table top in the center is also brighter.

One of the problems with 3D scanning using point clouds is that the laser light jumps off of transparent glass surfaces and curtain walls that are not solid walls, making it difficult to express a sense of depth, for example, and the data is too heavy to be easily processed on the spot. However, there is no doubt that the digitalization of interior and exterior spaces has made great technological strides.

SIMULATION

Chapter 2
5

ユビキタス情報による人間の行動軌跡
1. 下北沢における行動軌跡を見える化する

概要

　現在では、QRコードを用いたネット環境は格段に進化し、誰でも世界中の情報を一瞬にして得られるようになった。しかし、まちづくりや観光などにおいては、そこに行かなければ得られない情報の提供、具体的にいうと、その場の空間がもっている雰囲気や、音、香りなどを体験できることが、ますます重要になってきている。この実験では、2005年の時点ではまだ開発途上であったRFID（Radio Frequency Identification: 無線周波数識別）というチップあるいはQRコードを使いながら、行動パターンの特徴を分析した（回遊動線については一章22頁参照）。

下北沢地区におけるデジタル情報と行動軌跡の実験

　東京・下北沢地域は、商店会振興組合が7つもあり、2次元的に商業領域が広がっていることが大きな特徴である。この実験では、世田谷区下北沢のまちを歩きながら情報を受信、配信、共有することで、今まで知らなかったまちの魅力を発見することを目的の1つとした。

【実験の方法と概要】

　GPS機能付き携帯電話を用いた情報提供サービスと情報収集サービスを並行して実施した。具体的なサービスは、＜エリア情報サービス＞と＜クチコミサービス＞の2種類である。どちらのサービスも情報源に利用者が近寄ると自動的に情報メールが配信されるようにセットした。

1）エリア情報サービス

　特定エリアに入るとその近辺の情報（店舗の情報やイベント情報など）が自動的に携帯電話に配信されるサービスである。情報の供給方式は、シャワー型（A-type）、場所指定型（B-type）、定時場所指定型（C-type）の3種類が検討された。

■ A-type：シャワー型
　時刻の設定により、指定時刻になると下北沢全域に情報が配信される仕組み。地域内のイベント情報がシャワーのようにふってくる。
■ B-type：場所指定型
　情報発信源から半径50m以内に入ると情報が配信される。
■ C-type：定時場所指定型
　指定時間外に近くを通過しても何も起きないが、ある特定の期間内にある場所に近づいたもののみに情報が供給される。

　A. 時間限定型（場所:×／時間:○）
　B. 場所限定型（場所:○／時間:×）
　C. 場所時間限定型（場所:○／時間:○）

A. 時間限定型（場所:×／時間:○）

指定時刻に情報がシャワーのようにふる

14:00、踏切が開くのをまっていると…

B. 場所限定型（場所:○／時間:×）

情報源の半径50m以内に入ると情報が配信される

ラーメン屋さんに近づくと…

C. 場所時間限定型（場所:○／時間:○）

B. 場所限定型に時間設定がついたもの。時間外に近くを通過しても何も情報は配信されない。

15:00〜15:30にこの道を歩いていると…

2）クチコミサービス

　自分の知っている情報や新しく発見した街の魅力をクチコミとして投稿することで、他のサービス利用者と情報の交換を可能とするインタラクティブ型の情報供給サービスである。情報の配信方式は、

b-typeと同様の場所指定型とする。なお、サービス利用者から投稿されたクチコミはシモキタスマップ（地域資源マップ）に蓄積され、一度クチコミ情報を投稿すると、後からその情報源を訪れたユーザーにクチコミ情報が送信される。

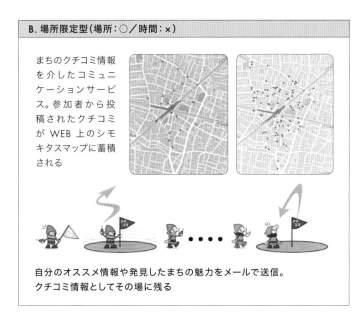

B. 場所限定型（場所：○／時間：×）

まちのクチコミ情報を介したコミュニケーションサービス。参加者から投稿されたクチコミがWEB上のシモキタスマップに蓄積される

自分のオススメ情報や発見したまちの魅力をメールで送信。クチコミ情報としてその場に残る

3）実験結果

　以下は、予め地域の商店主に依頼したサービス地点のリストと場所を示している。この半径50ｍの円内に入ると、その領域におけるサービス情報が自動的に伝えられるようにセットされた（Photo1）（Fig.1）。

（Photo1）ユビキタス実験の風景

11月23日の情報供給場所とタイプ

情報供給のタイプ
Total：20
A-type：01
B-type：12
C-type：07

（Fig.1）予めセットされたサービス地点の場所および供給内容のリスト
地元の商店に、ディスカントなどの特定のサービスを依頼した
I_a_01地点では、餅つきイベントを企画している

　以下は、実験の結果の行動軌跡を重ねたものであるが、行動軌跡が異なったタイプの半径50ｍの円に引き寄せられていることが分かる（Fig.2,3）。

11月23日の行動軌跡の重合　　GPS 各行動軌跡

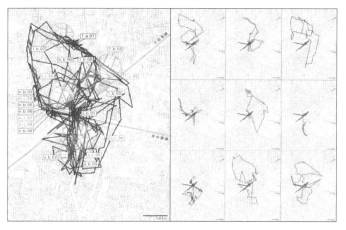

（Fig.2）ある実験日（11月23日）の行動軌跡を重合したパターン
人々の行動が、幾つかの結節点（ノード）を必ず通過している
また、普段なにもない場所で行われた餅つきイベントに多くの人が訪問をしていたことが分かる

情報供給と行動パターン

	23日(水)	24日(木)	25日(金)	26日(土)
情報供給状況				
行動経路図				

（Fig.3）日々異なったサービス地点の配置と行動軌跡のパターンの比較
クチコミによって日々異なる半径50ｍの円のパターンに
行動軌跡が引き寄せられていることが分かる

［対象者］約70名
［情報総数］91件（実験期間中の投稿総数）

　シャワー型による情報伝達は、インターネットで全国から取得できる情報と等価であるが、場所限定型の情報伝達は、その場に行かない限り得られない情報であるので、明らかに情報に引き寄せられて行動する傾向が見られた。また、クチコミによる参加型の情報提供により、さらに未知の情報が得られることになり、普段行かない場所へ行きたくなるようなインセンティブが働いていることも分かった。アンケート結果からも、実験による体験について、好意的な回答が得られた（Fig.4）。

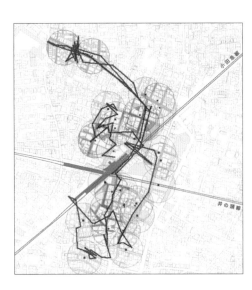

（Fig.4）サービス地点の
配置に影響された
行動軌跡のパターン事例

〈 CHAPTER 2, 5-1 〉 UBIQUITOUS INFORMATION AND
                      HUMAN ACTIVITY TRAJECTORY

Visualization of Human Activeity Trajectory in Shimokitazawa

SUMMARY

The use of QR codes online has evolved dramatically, and anyone can obtain information from around the world instantly. However, in urban development and tourism, it is becoming increasingly important to provide information that cannot be obtained without going there, or more specifically, to be able to experience the atmosphere, sounds, and smells of a particular place. In this experiment, we used RFID (Radio Frequency Identification) chips or QR codes, which were still under development in 2005, to analyze the characteristics of behavioral patterns.

An Experiment on Digital Information and Action Trajectories in the Shimokitazawa Area The Shimokitazawa area of Tokyo is characterized by the presence of seven shopping street associations and a two-dimensionally expanded commercial area. One of the objectives of this experiment was to discover the charms of the area by receiving, distributing, and sharing information while walking around Shimo-Kitazawa district.

Method and outline of the experiment

An information provision service and an information collection service were conducted in parallel using GPS-equipped cell phones. Two specific services were provided: <area information service> and <word-of-mouth service>. Both services were established so that users would automatically receive a digital information when they approached the information source.

1. Area Information Service

This service automatically delivers information (including store and event information) about the vicinity of a specific area to a cell phone when the user enters the area. Three types of information delivery methods were considered: Shower type (a-type), Location-specified type (b-type), and Fixed-time location-specified type (c-type).

a-type: Shower type

A system in which information is distributed to the entire Shimokitazawa area at a specified time by setting the time. Information on events in the area was to be distributed like a shower.

b-type: Location-specific type

Information is distributed when people enter within a 50-meter radius of the information source.

c-type: Regularly scheduled location-specified type

Nothing happens if you pass by the location outside of the designated time period, and information is provided only to those who approach the location within a certain period of time.

2. word-of-mouth service

This is an interactive type of information service that allows users to exchange information with other service users by posting word-of-mouth messages about information they know and newly discovered attractions of in the city. The information distribution method is the location-specific type similar to b-type. Word-of-mouth information posted by service users is stored in the SHIMOKITAS map (regional resource map), and once word-of-mouth information is posted, it is sent to users who visit the information source later.

1. Area Information Service

A service that automatically delivers information (e.g., store advertisements and event information) about the vicinity of a specific area to cell phones once the user enters that area. People can obtain information based on their location.
A. Time-limited type (Location: X / Time: ◯)
B. Location limited type (Location: ◯ / Time: X)
C. Location time limited type (Location: ◯ / Time: ◯)

A. Time-limited type (Location: X / Time: ◯)
Information is sprinkled like a shower at a specified time.

B. Location limited type (Location: ◯ / Time: X)
Information is delivered when you enter within a 50m radius of the information source.

C. Location and time limited type (Location: ◯ / Time: ◯)
B. Location limited type with time setting.
No information is delivered even if you pass nearby outside of the time.

2. Word-of-mouth service

B. Location limited type (Location: ◯ / Time: X)
Communication service through word-of-mouth information about the town. Word-of-mouth information posted by participants is stored on the SHIMOKITAS Map on the Web.

Participants can send their recommendations and other town attractions they have discovered by digital communication. The information will remain as word-of-mouth information.

Results of the experiment

The following is a list of service points and locations requested in advance from local store owners. The system was established so that once a visitor entered a circle with a radius of 50 meters, information about services in that area was automatically conveyed to the visitor. (Fig.1)

Local merchants were requested to provide specific services such as discernment; at location l_a_01, a mochi pounding event was planned.

People's behavior always passed through several nodes. It can also be seen that many people visited the mochi pounding event, which was held in a place where there is usually nothing to do.

The results show that the action trajectory is drawn to the pattern of a circle with a radius of 50m, which differs daily by word-of-mouth communication.

Subjects: Approximately 70 persons
Total number of information: 91 (total number of postings during the experiment)

Shower-type iInformation transmission by shower type is equivalent to information that can be obtained from all over the country via the Internet, but location-specific information transmission is information that cannot be obtained unless one goes to the location, so there was a clear tendency to be drawn to the information and act upon it. In addition, it was also found that the provision of participatory information by word-of-mouth communication further led to the acquisition of unknown information, which in turn created an incentive to go to places one would not normally go. The results of the questionnaire also showed that the participants responded favorably to their experiences in the experiment.

Fig.1　l ist of pre-set service point locations and supply details
Fig.2　Overlaid pattern of action trajectories on one experimental day (November 23)
Fig.3　Comparison of daily different service point locations and behavior trajectory patterns
Fig.4　Example of action trajectory patterns influenced by the placement of service locations

Chapter 2
5
ユビキタス情報と人間の行動軌跡
2. 高梁市における行動軌跡を見える化する

概要

　70頁の下北沢に対して、地方都市における情報の提供と行動パターンについて、岡山県高梁市という城下町で同様の実験を実施した。重要文化財である備中松山城や、頼久寺などさまざまな景観資源を抱えているため、観光客がどのように市内の観光資源を巡るかを探った。初めて観光地を訪れた者にとって興味深い観光情報が与えられれば、行動が活発化し、満足感を得られることが期待される。

高梁市におけるデジタル情報と行動軌跡の実験

　地方都市において、地域の隠れた資源を生かした「地域観光」を促進することは重要な施策である。特に最近では従来の「通過型観光」ではなく、長期滞在型の「体験型観光」や「発見型観光」に観光メニューが変わりつつあるため、なおさら外部から訪れた人たちがいつでもどこでも充分な地域情報を得られる環境を構築したい。ユビキタス技術にはさまざまな種類があるが、ここでは地域情報の提供と同時に、人の行動をGPS装置によりトレースすることで、街なかの回遊性を促進する仕掛けづくりを探る。

高梁における社会実験の概要

1. 予備実験 (2009年8月)

　11月に行う本実験に先駆けて、予備実験を実施した。これは、携帯電話のバーコードリーダーを利用して各観光資源に設置した「UcodeQR」を読み取ることで、観光情報、ロケ地情報、生活情報、周辺情報などを自由に取得できるようにし、それを利用して観光してもらうことの可能性を確認することを目的とした。また、まちの人の協力を得てGPS装置を携帯してもらい、行動トレースの可能を探った。

2. 本実験 (2009年11月)

　本実験では、予備実験の情報メニューに飲食店や物販店等の情報を加えると共に、「ポイント制」を新たに採用することで、より積極的に訪問者や生活者に利用してもらうためのインセンティブシステムの可能性を検証する。

ポイント制：「UcodeQR」を読み取ることで、スタート地点からそのポスターの距離に応じたポイントを取得できるようにする。遠くへ行くほど高いポイントをもらえるなど、取得ポイントに差を付けることで、より多彩な人の動きが期待できる。また、例えばこのポイントを地域通貨と交換し、それを地元商店街で利用できるなど、工夫次第で色々な

利用方法が可能である（Fig.1,2）。

(Fig.1) 予備実験の流れ

(Fig.2)「UcodeQR」の具体的な配備方法の提示

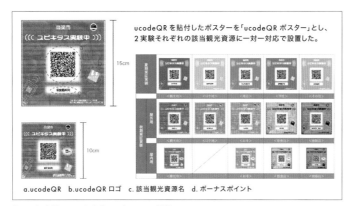

ucodeQRを貼付したポスターを「ucodeQRポスター」とし、2実験それぞれの該当観光資源に一対一対応で設置した。

15cm

10cm

a.ucodeQR　b.ucodeQRロゴ　c.該当観光資源名　d.ボーナスポイント

（Fig.3）色違いで示した「UcodeQR」の種類
ロケ地、寺社、飲食店などのQRコードを色分けして作成した

【実験の実施方法】

1）街中にある主要な観光拠点や映画ロケ地などに「UcodeQR」の載ったポスターを設置した。この「UcodeQR」は、その場所に応じた様々な情報（観光情報・シネマ情報・生活情報・グルメ情報・関連情報など）を取得する為に、携帯電話から情報ネットワークへアクセスするコードとなる。

2）実験参加者や観光客には一定時間、携帯電話（原則として本人の物）とGPS端末を持って街中を歩き回って頂き、各所の「UcodeQR」から必要な情報を取得してもらう。

3）実験の中で、提供する情報の違い（誘導する・しないなど）、ポスターの位置によるポイント数の変化などで、観光客の動きにどのような変化が起こるのかを観察し、検証する。この「人の動き」については、持ち歩いて頂いたGPS端末で取得した情報を利用しトレースする。また、実験後にアンケートを実施し、実験全体の印象や「UcodeQR」の使用感について調査する（Fig3,4,5）。

【実験の成果】

1. 予備実験の成果

予備実験では、街なかの中心部の人の動きや、行動パターンを詳細にトレースできることが確認された。また、自転車による行動エリアの拡大を具体的に確認することができた。

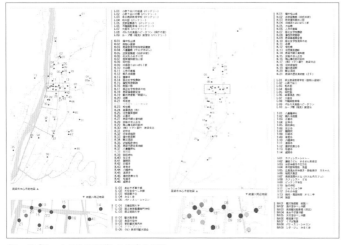

（Fig.4）「UcodeQR」を配備した市内の場所

提供情報について

（Fig.5）本実験では予備実験のスレッドにグルメ情報を追加した

2. 本実験の成果

本実験では、ポイント制の導入の有無により、行動パターンに明らかに違いが表れた。また、それぞれのスレッドに対応した行動パターンが明確に確認できた（Fig6,7,8）。

1）ポイントシステムON/OFFによる回遊行動

街なかでは、ポイントON時の方がより細かく広範に回遊行動を行っていることが分かる。また、紺屋川美観地区や栄町商店街を見てみると、ポイントOFF時はぽつぽつと「UcodeQR」の読み取りが行われているのに対し、ポイントON時は近隣の「UcodeQR」がまとまって読み取られていることが見て取れる。ここでは、ポイントを得る行為がもう一歩足を伸ばすきっかけになったと考えられる。またポイントを集めながら観光するというプラスアルファの要素が加わったことで、より細かく「UcodeQR」の読み取りを行うようになったと考えられる。

2）読み取られた「UcodeQR」と読取回数に関する比較

全体的に見ると、ポイントOFF時には主要な観光資源以外は読み取られていないのに対し、ポイントON時には細かい観光資源も1〜4回の読み取られている。また、観光名所に関してポイントON時、明らかに読み取り回数が増えている。

3）考察

全体にポイントが加算されるという"ゲーム性"が行為を誘発したと考

GPSによるトレーサビリティ　　　　　6日 10:00〜12:00　学生 男

学生（初めて訪れた人）は、国道沿いを歩くなど、大回りする傾向が見受けられる。今回の実験ではルート案内を行っていないので、それが顕著に反映されたと考えられる。

頼久寺、寿覚寺、道源寺など、寺関連の施設を順にめぐっており、寺の「関連情報」が有効に使われていると考えられる。

（Fig.6）予備実験の成果の一例
紺屋川という情報が多い場所での行動回遊性が見られた

えられる。また、何らかの興味(今回のポイント加算、歴史への興味等)がなければ観光客が足を伸ばさない可能性のある観光資源に対しても、ポイントシステムは有効に働くことが確認された。

(Fig.7)ポイントONの場合の
読み取られた「UcodeQR」の
回数と行動軌跡

(Fig.8)ポイントOFFの場合の
読み取られた「UcodeQR」の
回数と行動軌跡

　この実験を通して、不特定多数の人の街なかにおける行動パターンの把握によって、ユビキタス技術が観光客にとって必要な情報提供のあり方を示したことや、人々の回遊や散策行動を促すことに有効に働くことが分かった。また、人々が都市空間で取得した情報収集をもとに移動した経路を把握できるので、広場等の公共空間計画や公共施設計画、またサイン計画など、まちづくりを推進していく際にも有効に利用できる。

CHAPTER 2, 5-2 　UBIQUITOUS INFORMATION AND
HUMAN ACTIVITY TRAJECTORY

Visualization of Human Activeity Trajectory in Takahashi City

SUMMARY

In response to Shimokitazawa on page70, we conducted a similar experiment on information provision and behavioral patterns in a regional city, Takahashi City, Okayama Prefecture, which is a castle town. Since the city has various scenic resources such as Bitchu Matsuyama Castle, an important cultural property, and Raikyuji Temple, we explored how tourists visit the city's tourist resources. It is expected that if first-time visitors are given interesting tourist information, their behavior patterns will become more active and they will feel more satisfied.

An Experiment of Digital Information and Action Trajectory in Takahashi City

In local cities, it is an important measure to promote "regional tourism" that takes advantage of hidden local resources. Recently, tourism menus have been changing from traditional "transit-type tourism" to "experience-type tourism" and "discovery-type tourism," where visitors stay for a long period of time. Amongst various ubiquitous technologies, we will focus on the mechanism to provide local information and ways to facilitate people's movements in the city by tracing them using GPS device.

Outline of the social experiment in Takahashi

1. Preliminary experiment (August 2009)

A preliminary experiment was conducted prior to the main experiment in November to confirm the possibility of reading "UcodeQR" installed at each tourist resource destinations using a cell phone, where visitors can freely obtain various tourist information, such as location, daily life, and neighborhood, for sightseeing. In addition, we asked the local residents to carry GPS devices to explore the possibility of tracing their movements.

2. main experiment (November 2009)

In this experiment, information on restaurants and stores was added to the menu of the preliminary experiment, and adopted a new "points system" as an incentive system to encourage visitors and consumers to use the system more actively.

Points system: Visitors are able to accumulate points based on the distance of the poster from the starting location by reading the "UcodeQR". We have differentiated the points that can be earned, for example, more points for farther distances, in order to see varied movements of people. These accumulated points can then be used for local shopping within the area.

Method of implementation of this experiment

(1) Posters with "UcodeQR" were placed at major tourist attractions and movie locations in the city. These codes were used to obtain various types of information such as sightseeing, c inema, and daily life etc. related to the area.

(2) Participants in the experiment and tourists are asked to walk around the city for a certain period of time with their own cell phone and a GPS terminal, and obtain necessary information from "UcodeQR" at various locations.

(3) During the experiment, we observed and verified how the movement of tourists changes according to the differences in the information provided and the number of points that can be earned depending on the location of the posters. The "movement of people" was traced using GPS terminals carried by the visitors, and a questionnaire was carried out to find participants' impression of the experiment, and how they felt about using the UcodeQR.

Results of the experiment

1. Results of the preliminary experiment

In the preliminary experiment, it was confirmed that the movement of people in the center of the city and their activity patterns can be traced in detail including expansion of bicycles areas.

2. Results of this experiment

In this experiment, a clear difference in the behavioral patterns was observed depending on whether a point system was introduced or not. In addition, the behavior patterns corresponding to each thread were clearly identified.

(1) Migration behavior by point system ON/OFF

In the city center, visitors make more detailed and extensive trips when the point system is turned on. In the Konyagawa Bikan Historical Area and Sakae-machi Shopping District, it can be seen that UcodeQRs are read one by one when points are turned off, while they are read in neighborhood clusters when points are turned on. Therefore, earning points is thought to have triggered the visitors to go one step further, as well as leading them to use the UcodeQR more frequently.

(2) Comparison of "ucodeQR" read and number of times read

Overall, while only major tourist attractions were read when points were turned off, detailed tourist attractions were read 1 to 4 times when points were turned on. In addition, the number of readings of tourist attractions clearly increased when points were turned on.

(3) Consideration

It is thought that the "gameplay" of having points triggered tourists' behavior. It was also confirmed that the point system works effectively for tourist resources where people might not visit without some interest.

behavior patterns of an unspecified number of people in the city, ubiquitous technology showed the way to provide necessary information for tourists, and that the system was effective in encouraging people to migrate and stroll around the city. Moreover, since we can identify the routes people traveled from the collected information, it can be effectively used to promote urban development, such as planning of public spaces, public facilities, and signages.

Fig.1　Preliminary experimental flow
Fig.2　Presentation of the specific deployment method of "UcodeQR"
Fig.3　Types of "UcodeQR" shown in different colors
Fig.4　Locations in the city where "UcodeQR" was deployed
Fig.5　Gourmet food information was added to the thread of the preliminary experiment in this experiment.
Fig. 6　An example of the results of the preliminary experiment
Fig.7　Number of "UcodeQR" read and behavior trajectory when points are turned on
Fig.8　Number of "UcodeQR" reads and the trajectory of the action when the point system is turned OFF.

都市における建物用途の「雑多性」と地域の特徴
建物用途の「雑多性」を見える化する

概要

　わが国においては、一般に用途地域によって建物用途が規制されているにもかかわらず、商業や近隣商業地域ではそれが緩和されているため、いわゆる"雑居ビル"と呼ばれる多用途な建物が市街地に多く存在する。ここでは、建物の上階まで多様な用途がテナントとして占有していることに着目し、ある街路空間の雰囲気が、街路に面する両側の建物用途の多様性によって影響を与えられるか（にじみ出し）を検証する。具体的には、東京都の各地区における多様な用途の集積を3Dで示し、目に見えない用途の集積が醸し出すエリアのアイデンティティを視覚化する（多用途性については一章24頁参照）。

異なる用途の間に存在する「差異」とは

　アメリカの作家・ジャーナリストであるジェーン・ジェイコブス（1916～2006年）は、ニューヨークの街並みに古い建物や新しい建物が混在し、建物の用途が多様であればあるほど楽しい居住環境が生まれると説いた。実際、1919年に日本国内で制定された用途地域制度は、住居地域・商業地域・工業地域の3用途に分類する単純な政策であったが、居住環境を守るという命題は正しかったものの、その後、日本中に均質な都市を生み、多様で雑多なまちが生まれにくい仕組みとなった。現在は13の用途地域にまで細分化されているが、その縦割的構造は殆ど変わっていない。

　建物の用途には「住居に良い影響を与える」というベクトルと「住居に悪影響を与える」という2つのベクトルが存在している。そのため、住居系の用途地域では建蔽率を抑え、建物の許容容積をできるだけ低く抑えているのに比べ、商業地や工業地域では、建物のボリュームを敷地いっぱいに建て、高密度になることを許容している。これらは相反する両極なベクトルであるので、ある建物群の用途多様性や雑多性の度合いを調べることにより、この2つの方向性の影響度を一種の物差しとして、地域がもつ特徴を示せると考える。なぜなら、異なった建物用途には、異なった属性の人々が出入りし、その街路にも独特の雰囲気を与えているからである。そこで、異なる用途の間には何らかの「差異」が存在することに着眼し、その「差異」の度合いによって、地域の雰囲気や界隈的特徴が影響を受けているという考えをもとにシミュレーションを進める。

「用途の差異」と「雑多性」
繁華街は「用途の差異」が顕著である

　都市において、「用途の差異」が顕著にみられる地域は、やはり渋谷・新宿といった賑わいのある駅前繁華街である。これらの商業界隈はその用途・景観の両面から「雑多な街」と形容され、それが界隈の特徴であ

ると言われてきた。一般に「雑多」という言葉は、いろいろなものが入り混じっている様子を意味し、差異の集積から生じる考え方である。たとえば、「用途の差異」が大きく「雑多」の度合いが強い地域は賑わい性が高く、逆に「雑多」の度合いが

（Photo1）繁華街における店舗看板
雑多性が高く、町のにぎやかさが醸し出されている

弱い地域は均質性が強くなると考えられる。ここでは、このように「用途の差異」から生じる界隈の特徴を地域の「雑多性」と定義し、さらに「雑多性」の強さの度合いを示すものとして「雑多度」という指標を用いる。表現方法としては、「雑多度が高い地域＝多用途性がある」、また反対に「雑多度が低い地域＝多用途性がない」ことになる。

　（Photo1）の看板は飲食、エステ、カラオケ、消費者金融、映画館、そして事務所とさまざまであり、一般に繁華街といわれる地区では、建物用途間の差異による「雑多度」は高い。

神田神保町における雑多性

　神田神保町は古本屋街で知られるが、スポーツ店や楽器店などが集中している界隈にも隣接している。この地区全体の建物の各階用途を綿密に調べ、3Dの分布図に表現し、1）鳥瞰的に見た建物用途分布図と、3）アイレベルから見た建物用途の分布図、2）実際に展開されている街並みを比較し、建物の雑多性と地域の雰囲気、都市景観について考える（Fig.1,2）。

　全体的に、垂直にさまざまな用途が入居しており、極めて雑多性が高い地区であることが分かった。この地区の"いわゆる"ペンシルビル群は、江戸時代の狭い町割りの上に重ねられた建物空間であるが、渋谷や新宿のような飲食系の雑居ビルではないので、外看板の数は少なく、都市景観的には落ち着いている傾向である。
靖国通り沿いには、古書店や新
刊書店が多く
並び、神田
すずら

（Fig.1）千代田区神田神保町地区における建物用途の雑多度を示す鳥観図

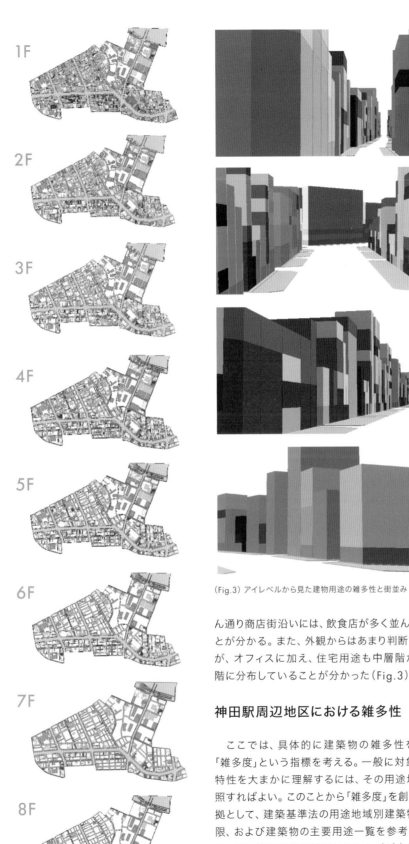

1F

2F

3F

4F

5F

6F

7F

8F

(Fig.2) 各階の建築用途分布図

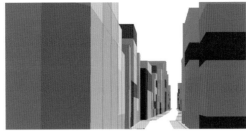
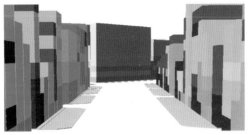
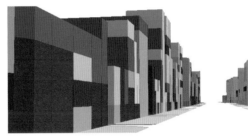
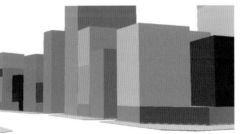

(Fig.3) アイレベルから見た建物用途の雑多性と街並み

ん通り商店街沿いには、飲食店が多く並んでいることが分かる。また、外観からはあまり判断できないが、オフィスに加え、住宅用途も中層階から上層階に分布していることが分かった（Fig.3）。

神田駅周辺地区における雑多性

　ここでは、具体的に建築物の雑多性を調査し、「雑多度」という指標を考える。一般に対象地域の特性を大まかに理解するには、その用途地域を参照すればよい。このことから「雑多度」を創出する根拠として、建築基準法の用途地域別建築物用途制限、および建築物の主要用途一覧を参考に作成する。特に神田駅周辺地区はほとんどが商業地域であることから、商業地域の建築物用途制限を参考にし、「雑多度」を創出する方法を考える（Fig.4）。
　まず、建築基準法の建築物用途制限の表を参考に、"「住む」環境に良い影響を与える用途"と"「住む」環境に悪い影響を与える用途"を両極としたグラデーションを作成する。この並び方に関する妥当性については、当然議論する余地があるが、ここでは、建築基準法を参考にしたケーススタディーとして位置付けておく。

(Fig.4) 神田駅周辺の対象地区
商業用途地域に指定されており、雑居ビルが多い

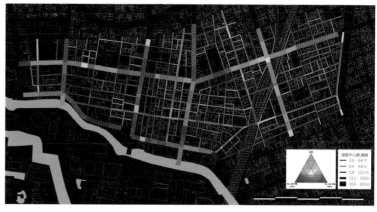

（Fig.10）道路両側の建物総合用途の平均による道路の「雑多度」の分布
白色に近いほど道路の両側に多用途の建物が集積しており
単色に近いほど単用途の建物が集積している

次に、グラデーション上にそれぞれの建物の用途を分類し、凡例を作成する。次に建物の各階の用途を調べ、凡例に沿って色付けを行い、各階の色を光の三原色のように色を混ぜる。建物が単用途の場合は純色となり、多種の用途が複合している建物は白で色付けがされる。たとえば、純緑色は住宅のみの用途の建物であり、黄色は住宅の用途と住宅にいい影響を与える用途の割合がそれぞれ50％ずつの建物となるようになる。建物をそれぞれ用途ごとに正確にかつ細かく色分けすることは難しいため、簡略化した方法をとる（Fig.5,6,7）。

住宅系	住宅	1
	教育施設 / 保育所 / 児童厚生施設 / 老人ホーム / 図書館 / ギャラリー / 美術館	2
	派出所 / 消防署 / 保育所	3
	大学・高等学校 / 病院	4
商業系	飲食店 / 物販店	5
	スポーツ施設 / サロン / 美容室等のサービス店	6
	オフィス・事務所 / 銀行 / 作業面積 150 m²以下の工場	7
	ホテル・旅館	8
	劇場・映画館等	9
	居酒屋・バー等の料理店 / カラオケボックス等	10
	キャバレー・クラブ等 / パチンコ	11
	風俗店	12
工業系	工場・供給施設・処理施設	13
その他	倉庫 / 駐車場	-

「住む」環境に良い影響を与える用途
「住む」環境に悪い影響を与える用途

（Fig.5）住む環境を軸にした建築用途のグラデーション図

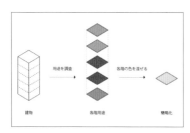

（Fig.6）各階の建築用途を総合し
総合的用途の雑多度を合成色で示す

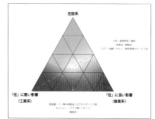

（Fig.7）各階の建築用途による色彩の合成を示すダイアグラム
建物用途の合成色が白色であるほど「雑多度」が高く、純色になるほど「雑多度」が低い

次に、対象地区の建物用途および各階用途を調べてマッピングし、建物用途分布図を作成する。この段階では、建物の敷地内の総合的用途の分布図を示している。このマッピング図は、総合用途である点が違うが、一般の建物用途図に近い（Fig.8）。

次に、各道路の両側にどのような用途が集積しているのかを考える。具体的には、（Fig.9）のように各道路に接触している建物用途の色を合成し、それを平均化する方法をとる。それにより各道路における用途の集積具合を表すことができる（Fig.10）。

渋谷と下北沢の雑多度

同様の考えにより、渋谷地区と下北沢地区の商業地区における建築用途の雑多度を調査する。これらの分析図では、階層によって用途が異なっているが、渋谷界隈では、駅周辺の建物の低層部に商業用途があり、上階にはオフィス用途が集中している。一方、下北沢界隈では、駅周辺の建物の上層部には住宅が点在し、後背地はほぼ住宅で埋められている（Fig.11,12）。

（Fig.8）敷地内の建物の総合用途をマッピングした地図

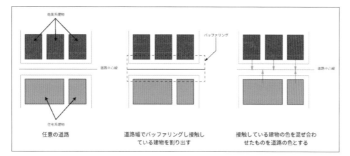

（Fig.9）道路に対して両側の建物総合用途を表出する方法
各道路に面している建物の総合用途の色を抽出し、それらを合成した色を道路に反映する

（Fig.11）渋谷駅周辺地域における建物用途図
JRと私有鉄道のターミナル駅でもあるため駅周辺にオフィス用途が多い

（Fig.12）下北沢駅周辺地域における建物用途図
私有鉄道が交差する駅であるが駅周辺から周縁部にかけて住宅用途が多い

これらの地区について、同様の「雑多度」を想定し、その度合いを高さで示す立体グラフを作成した（Fig.13,14）。

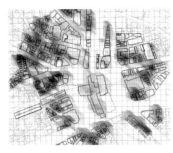
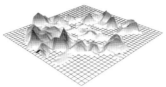

（Fig.13）渋谷駅周辺地域における建物用途の「雑多度」の分布
センター街を中心に、建物用途の「雑多度」が高いことが分かる

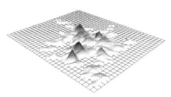

（Fig.14）下北沢駅周辺地域における建物用途の「雑多度」の分布
北側の一番街商店街および南口商店街の建物用途の「雑多度」が高いことが分かる

これらのグラフは、私たちの感覚にも近いものがあり、「雑多度」の高い地域には、さまざまな業態の生業が集まり、それに従って異なった種類の人たちが出入りするため、ジェーン・ジェイコブスの言う「建物用途の多様性」がまちの賑わいを生む、という言説をある程度裏付けることができた。

CHAPTER 2, 6

Visualization of the "Miscellaneousness" of Building Uses

SUMMARY

In Japan, building use is generally regulated by zoning, but this has gradually eased in commercial and neighborhood commercial districts, resulting in the presence of many so-called "multi-purpose buildings (Zakkyo buildings)" in urban areas, with a variety of programs occupying the building floors as tenants. In this study, we examine whether the atmosphere of the street space can be influenced by the diversity of building use on both sides facing the street by visualizing the hidden various programs taking Kanda Jimbocho, Kanda Station area, Shibuya, Shimokitazawa, and Ginza districts in Tokyo as case studies. However, here we will take Kanda station area as an example to explain the "degree of miscellaneousness" concept in detail.

Degree of Miscellaneousness: Difference in Use and "Miscellaneousness"

Generally, in urban areas, "differences in use" are most noticeable in bustling downtown areas in front of train stations, such as Shibuya and Shinjuku. These commercial districts have been described as "miscellaneous towns" in terms of both use and townscape. Here, we define "miscellaneousness" as the characteristics of a neighborhood resulting from the "difference in use," and use the "degree of miscellaneousness" index to indicate its strength. In other words, an area with a high degree of miscellaneousness = multi-use, while an area with a low degree of miscellaneousness = no multi-use.

The signs in Photo 1 are for restaurants, esthetic salons, karaoke, consumer finance, movie theaters, and offices. In general, the "degree of miscellaneousness" is high in areas that are known as downtown.

"Degree of miscellaneousness" in the Kanda Station Area

In general, to obtain a rough understanding of the characteristics of a subject area, it is sufficient to refer to its zoning district. Based on this, the "degree of miscellaneousness" is created by referring to the Building Standards Law's restrictions on building use by zoning district and the list of major uses of buildings. In particular, since most of the area around Kanda Station is a commercial district, we will consider how to create "miscellaneousness" by referring to the building use restrictions in the commercial district.

First, referring to the table of building use restrictions in the Building Standards Law, we distinguish between "uses that have a positive impact on the living environment" and "uses that have a negative impact on the living environment."

Next, the uses of each building are classified on the gradient and a legend is created, which is applied to examine the use of each floor of the building. then by mixing the colors of each floor like the three primary colors of light. If the building has a single use, it is a pure color, and if the building has a combination of many different uses, it is colored white.

Next, the building use and each floor use in the subject district are examined and mapped to create a building use distribution map.

Then, in order to consider what uses are clustered on both sides of each street, the colors of building uses in contact with each road are combined and averaged. This allows us to express the degree of concentration of uses on each street.

Miscellaneousness in Shibuya and Shimokitazawa

The same idea is applied to investigate the degree of building use in the commercial districts of Shibuya and Shimokitazawa. In the Shibuya neighborhood, commercial uses are concentrated in the lower levels of buildings around the station, while office uses are concentrated on the upper floors. In the Shimokitazawa neighborhood, however, the upper floors of buildings around the station are dotted with residential buildings, and the hinterland is almost entirely filled with residential buildings.

Photo 1 Store signage in the downtown area
Fig.1 Bird's eye view of the Kanda Jimbocho district in Chiyoda-ku, Tokyo, showing the variety of building use
Fig.2 Distribution of building uses on each floor
Fig.3 Miscellaneous building uses and townscape seen from eye level
Fig.4 Target district around Kanda Station
Fig.5 Gradient diagram of building use based on living environment
Fig.6 The building uses on each floor are synthesized, and the degree of miscellaneousness of the overall uses is indicated by the composite color
Fig.7 Diagram showing the synthesis of colors by building use on each floor
Fig.8 Map of overall building uses on the site
Fig.9 Method of showing the overall use of buildings on both sides of a road
Fig.10 Distribution of "miscellaneousness" of a road based on the average of the comprehensive use of buildings on both sides of the road
Fig.11 Building use map in the area Surrounding Shibuya Station
Fig.12 Building use in the Shimokitazawa Station area
Fig.13 Distribution of "degree of miscellaneous" of building uses in the area surrounding Shibuya Station
Fig.14 Distribution of "miscellaneous" building uses in the Shimokitazawa Station area

SIMULATION

概要

建物の形状や機能によって分類する方法をビルディングタイプに分類するというが、これには、"学校、図書館などの機能別ビルディングタイプ"と"歴史的建物、近代建築などの形態的特徴によるビルディングタイプ"がある、ここでは、後者の類型がどのように分布しているかを視覚化する方法を検討する（GISについては一章26頁参照）。

都市景観の「まとまり」を見える化する

1章の視覚による認知（16頁）で述べた「まとまり」について改めて考えてみよる。「まとまり」とは多様性のある都市景観の中で、ある共通の特徴で個々の建築が1つの集合体としてまとまって認識される状態を言い、その共通性が秩序ある都市景観をつくりだす鍵になっている。極端に言えば、「まとまり」は都市景観の最小の単位として存在し、逆に都市景観は「まとまり」の集積として形成されていると考えられる。しかし、現代日本の街並みは西洋のそれと比較すると、景観としての「まとまり」が稀薄であり、僅かな形態的共通点さえも見つけにくい混沌的状況となっている。この状況を改善するためには、都市景観という漠然とした事象を誰の目にも共通に分かるように示すことが重要である。また、抽象的な事象を具体的なかたちとして表示することで、複雑化した現代都市をより単純化して考察することが可能になると思われる。ここでは、都市がどのような「まとまり」によって構成されているのかを3Dグラフィックスを用いて視覚化するための方法論を考える。最終的には、現代都市景観における問題点や、これから先の都市デザインの方向性を視覚的に導き出すことを目的にしている。

しかし、「まとまり」と言っても、その存在を一概に示すことは難しい。特に歴史的街並と近代建築が共存する混在型都市においては、都市の中でどれだけの「まとまり」があるのかを感じるのさえ困難である。そこで要素がどれだけ集積しているのかを具体的に示すものとして、ここでは密度に着目する。実際の都市景観において構成要素の集積度合いは重要なポイントである。ここでは形態的密度に着目することで建物の特徴的な形態的要素の分布状況を明らかにし、その集積度の大きさから、都市景観における「まとまり」の強さを視覚化する。

形態的要素の類型化

まず、形態的要素として抽出する建築の類型化を行う。ここでは、住居に焦点を当てて考察するため、住居形式という点から細かい類型化を試みる。対象地域である岡山県高梁市おける住居形式は、大きく「町家形式・屋敷形式・郊外住宅形式・集合住宅形式」の4種類に分類することができる。しかし建物形状だけの分類では、歴史的様式によって建てられている住居と、近代様式で建てられている住居との区

別ができないため、各形式を「伝統継承型・非伝統継承型」の2つに分類し、それを掛け合わせた8種類についての要素を抽出する（Fig.1）。

住居形式の類型

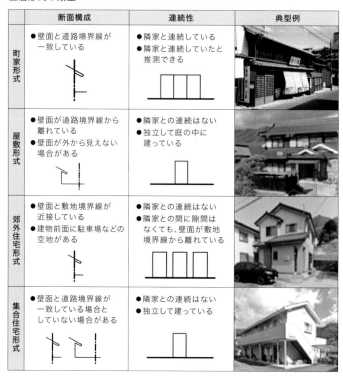

（Fig.1）高梁市における形状の違いによる住居形式の類型を示す
これに歴史的様式、近代様式をかけ合わせ、8種類のマトリックスを作成する

集積密度グラフによる「まとまり」の視覚化

ここでは、具体的に形態的要素の密度を視覚化する方法論について考える。まず抽出する要素について、その要素がどのように分布しているのかを調査する必要があるが、分布の様子をあらわす手段として、最もよく用いられる方法はマッピングによる「分布図」である。これは要素の分布の状況を二次元の図として表したものだが、これでは分散・集中の様子を知ることはできても、その集積度の差を見ることはできない。そこで、任意の点からある一定範囲（r）におけるある要素の分布密度を、その点の集積密度として計算する（Fig.2）。次に、集積密度の強さを視覚化するため、都市空間をグリッド状に分割した後、その交点における集積密度を高さで表現し、起伏のついた面状のグラフを生成する。これを「集積密度グラフ」と呼ぶことにする（Fig.3）。そしてこれにより異なった住居形式の「まとまり」の強さの視覚化を試みる。

高梁市におけるグリッド分割は、20m×20mを単位スケールとした。それはこの都市における町割の基本が町家形式の住居であり、短冊形

をしている町家一軒分の奥行きがおよそ20mであることから、20m
角のグリッド分けすることが適切であると判断したためである。また
半径何メートルにおける集積度を求めるのかという「r」の値について
は40mとした。それは認識実験により得られた住居群が交差点から
概ね40m以内の範囲に存在しているという点と、グリッド分割をする
にあたって基本とした町家形式の住居の奥行きが20mであり、主軸を
形成している道路間の距離が町家二軒分であることを根拠とした。

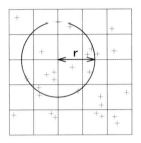

(Fig.2)集積密度
任意の点から一定範囲(r)における
要素の密度を測定する

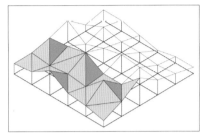

(Fig.3)集積密度グラフの例
一定間隔のグリッドの交点上に密度の
高さを表現する

フィールドワークによるプロットの方法

　高梁市における形態的要素
の集積密度について、まずそれ
らの要素の分布状況を調査し
た。これはフィールドワークに
あたるもので、すべての建築を
ビデオに収め、そのビデオを見
て分類した。そして判別した要

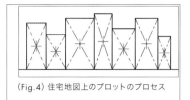

(Fig.4)住宅地図上のプロットのプロセス

素について地図上にプロットしていくプロセスでは、住宅地図を使用
し、その敷地の中心に点を打った(Fig.4)。下に、代表的な住居形式の
集積密度グラフを個別に示す(Fig.5)。

　次にこれらの集積密度グラフを重ねわせたグラフを作成した。1つ
目は、伝統継承型住居形式の集積密度グラフを重ね合わせた図であ
る(Fig.6)。高梁市の市内は城下町としての町家の通りと武家屋敷の
通りが明確に分かれているが、その状況がこの図にも明解に表れてい
ることが分かる。また、同形式の集積の度合いがグラフの高さに良く

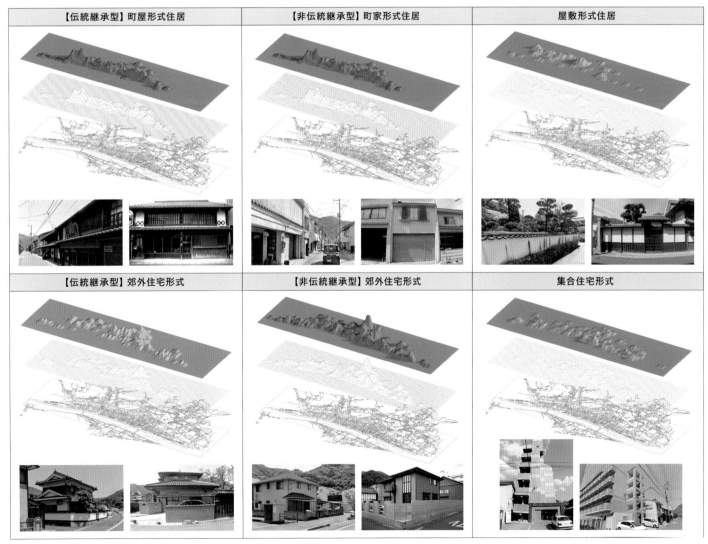

(Fig.5)住宅地図上のプロットのプロセス　代表的な住居形式の集積密度グラフを個別的に示している

現れ、目視で認識した街並み景観と符合していることが分かった。

　（Fig.7）は全類型の住宅形式の集積密度グラフを重ね合わせたものである。この図からは、郊外住居形式が山裾や周辺部に集中して開発されていることが分かる。

　以上の考察から、集積密度グラフが、私たちが日ごろ都市景観から感じている印象に近い状況を示していることが分かった。それと同時に、あまりに多様な建物形式が混在する都市景観には「まとまり」が感じられず、ある程度、同型式の建物が集積していることが、安定的な都市景観を構成することが明解になった。この知見は、新しい都市開発や景観の修復時の大きな指針になると思われる。

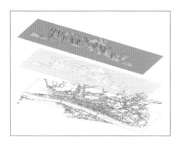

（Fig.6）伝統継承型住居形式の
集積密度グラフの重ね合わせ図

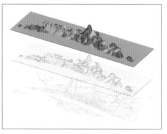

（Fig.7）全類型住宅形式の
集積密度グラフの重ね合わせ図

CHAPTER 2, 7-1　URBAN ANALYSIS BY USING GIS

Visualization of the Degree of Concentration of Building Type

SUMMARY

The method through which buildings are classified by form and function is called building type classification, which includes "building types by function, such as schools and libraries," and "building types by morphological characteristics, such as historical buildings and modern buildings." In this chapter, we examine how to visualize the distribution of the latter type.

Visualizing the "cohesiveness" of the urban landscape

Let us consider again the "cohesion" mentioned in Chapter 1, "Visual Recognition" (p.16). "Cohesion" refers to the state in which individual architectural structures are recognized as a single group with certain common characteristics in a diverse urban landscape. This commonality is the key to creating an orderly urban landscape. In extreme terms, "cohesion" exists as the smallest unit of the urban landscape, and conversely, the urban landscape is formed as an accumulation of "cohesion." However, compared to Western landscapes, the modern Japanese cityscape has little "cohesion," and even the slightest morphological similarities are difficult to find. In order to improve this situation, it is important to present the vague phenomenon of urban landscapes in an easily understandable manner. Moreover, by displaying abstract events in concrete form, it becomes possible to consider the increasingly complex modern city in a simplified manner. In this section, we consider a methodology for visualizing what kind of "cohesion" a city is composed of using 3D graphics. The ultimate goal is to visually derive the problems in contemporary urban landscapes and the future direction of urban design.

Since it is difficult to understand how much "cohesion" exists within the city in a mixed city in which historical streetscapes and modern architecture coexist, here we focus on density as a concrete indicator of the degree to which elements are clustered together. In the actual urban landscape, the degree of concentration of the constituent elements is an important point. By focusing on morphological density, we can clarify the distribution of characteristic morphological elements of buildings and visualize the strength of "cohesion" in the urban landscape based on the degree of concentration of these elements.

Categorization of morphological elements

First, we will categorize the architecture to be extracted as morphological elements. Since the focus of this study is on dwellings, we will attempt a detailed categorization in terms of residential form. In Takahashi City, Okayama Prefecture, the target area, the residential style can be broadly classified into four types: "machiya," "mansion," "suburban residence," and "collective housing." However, since it is impossible to distinguish between dwellings built in the historical style and those built in the modern style by classifying only building shapes, each type is classified into two categories – traditional and non-traditional – and elements are extracted for eight types by multiplying these two categories.

Visualization of "cohesion" using a density graph

In this section, we consider a method for visualizing the density of specific morphological elements using a "distribution map" based on mapping. This is a two-dimensional diagram of the distribution of the elements. However, although it can show the dispersion and concentration, it cannot show the difference in the degree of concentration. Therefore, the distribution density of an element in a certain range (r) from an arbitrary point is calculated as the accumulation density of that point (Fig. 2). Next, in order to visualize the strength of the density, the urban space is divided into a grid and the density at the intersection points is expressed in terms of height to generate an undulating surface graph. This is called an "agglomeration density graph" (Fig. 3).

In Takahashi City, the unit scale of grid division is 20m x 20m. This is because the town layout in this city is based on machiya-style houses, and the depth of a single strip-shaped machiya house is approximately 20 meters. The value of "r," which is the number of meters in radius, was set to 40 meters. This was based on the fact that the dwellings obtained in the recognition experiment were generally located within 40 meters of intersections, that the depth of the townhouse-style dwellings used as the basis for the grid division was 20 meters, and that the distance between the roads forming the main axis of the grid was two townhouses.

Fieldwork plotting method

To determine the density of morphological elements in Takahashi City, we first investigated the distribution of these elements. This was done through fieldwork, in which all the buildings were videotaped and classified. The identified elements were plotted using a residential map and dots were placed in the center of the site (Fig. 4).

Next, these density graphs were superimposed on each other, the first being the superimposed graph of the density of traditional residential types (Fig. 6). The city of Takahashi has a clear distinction between the streets of the castle town and the streets of samurai residences, which is clearly shown in this figure. In addition, the degree of accumulation of the same form is well represented by the height of the graph, which is consistent with the visually recognized townscape landscape.

Fig. 7 is a superimposition of the density graphs for all housing types. This figure shows that suburban residential types are concentrated in the foothills and peripheral areas.

From the above discussion, it is clear that the density graph shows a situation similar to the impression we usually get from the urban landscape. At the same time, it has become clear that an urban landscape with too many different types of buildings is not "cohesive" and that a certain degree of concentration of buildings of the same type constitutes a stable urban landscape. This finding will be a major guideline for new urban developments and landscape restoration.

Fig.1　Housing Type
　　　　The table shows the types of dwellings in Takahashi City according to their shapes
　　　　The matrix of eight types is created by multiplying the historical and modern styles
Fig.2　Area of accumulation density
　　　　at a fixed range (r) from a given point
Fig.3　Example of an accumulation density graph
　　　　Density is expressed on the intersection of grids at regular intervals
Fig.4 & 5　Plotting process on a residential map
Fig.6　Overlaid graph of the density of traditional residential types
Fig.7　Overlaid graph of the density of all housing types

Chapter 2 / 7

GIS による都市分析
2. 都市のグレインを見える化する

概要

ここでは、岡山県高梁市で試みた集積密度グラフ（80頁）をさらに進化させ、都市空間における建築（粒子）の集積のスケールと密度に着目し、その都市空間の形態的特徴を示す。その指標として、「グレイン（grain）分布図」を作成する理論を紹介する。この方法論の有効性を探るため、特徴的な4都市（歴史的都市 - 岡山県高梁市、地方中核都市 - 兵庫県姫路市、高密都市 - 東京都渋谷区、新宿区）について都市構造の視覚化を試みる。さらにこの視覚化した「グレイン分布図」を用いてそれぞれの都市の構成の仕方、都市の抱えている問題点、都市のもつ個性などについて「まとまり」的見地から考察する。

建築の集合体である"まとまり（グレイン）"

集落や都市は、複数の人間を内包する生活圏のための集合体として形成されてきた。そのため、粒子としての建築物は単独では存在せず、何らかの集合形態をとることとなった。古い集落や都市では、それは道に囲まれた「街区」として形成された。街区は古くから都市と建築の中間領域として存在し、都市の計画的単位を担ってきたが、わが国では戦後の高度成長以後、大型建築や超高層建築などが急増し、こうした粒子の著しいスケール的変化に対応できず、過去のように一定の秩序をもった街区とは異なった状態に変化させられてしまった。一般的に、街路パターンの形態的変化は少ないため、街区の形態は長年変化していないのに対し、街区の構成要素である粒子は時代の経過、社会経済の変動に対応して、その機能やスケールに変化を起こし、街区と粒子の間には大きなギャップが生じてきたと言えよう。

現在ある都市の計画的枠組みとして、地区計画やスーパーブロックなども挙げられるが、スケールの違いこそあるにしても、基本的な考え方は街区と相違はない。おそらく重要であるのは"現代の都市粒子を考えることのできる"計画的枠組みとその方法論である。ここで都市と建築の中間領域を担う新しい計画的枠組みの1つとして、「グレイン」のスケールを考えてみる。「グレイン」とは、現代における同質的都市粒子の集合体を意味しており、街区くらいの中間的スケール（メゾスケール）で都市構造を考えることにつながる。つまり、「グレイン」とは一様なスケールをもった建築物という粒子が一様な密度で集合したものであり、都市とは多様な「グレイン」の集合体としてとらえなおすことができる。

1950年代に、アメリカの都市学者であるケビン・リンチが提起したImageability（都市のイメージしやすさ）は、多くの人が共通の「まとまり」を知覚したときにはじめて、その都市の個性が認識されるということであった。この"イメージしやすさ"をもった「まとまり」は、都市において粒子の集合体である「グレイン」の明快な形態、もしくは分かりやすい分布状態といえるであろう。ここでは、いくつかの都市における粒

子の集積の仕方をシミュレーションし、都市の個性を「グレイン」的概念から考えてみる。

具体的に都市空間における粒子の集積の仕方を、そのスケールと密度に着目することで都市構造の視覚化を試みる。この視覚化した図では、都市の任意の点からある一定範囲内に分布した同質的粒子の密度をその点の高さの係数として与える。この高さ表現により、同質的粒子による一定密度によるまとまりである「グレイン」は、さまざまなタイプの山（突起）を形成する。この山の形態は粒子の分布状態によって変化し、山の高さは粒子間の密度の大きさに比例する。(Fig.1)のグレインA、グレインBは同質性粒子の2つの異なる密度によるまとまりを視覚化したもので、(Fig.2)のグレインC、グレインD、グレインEは3種類の異質性粒子の同一密度によるまとまりを視覚化したものである。

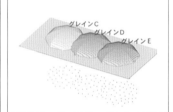

(Fig.1)密度の違いによるグレイン分布図の高さの違い

(Fig.2)異種の粒子によるグレイン分布図の重複のイメージ

ここでは、この粒子概念を導入、拡大定義し、建築物（粒子）のスケールと密度に着目することで、複雑化した現代の都市構造の視覚化を試みる。さらにこの視覚化した図（グレイン分布図）を用いて、建築の集合体である"まとまり（グレイン）"に着目しながら、各都市が持つ固有の構造を考えてみることにする（Fig.3）。

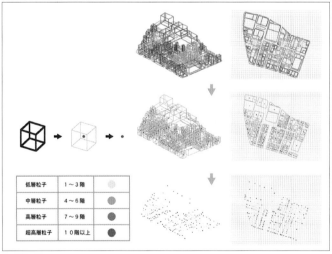

低層粒子	1〜3階	
中層粒子	4〜6階	
高層粒子	7〜9階	
超高層粒子	10階以上	

(Fig.3) 粒子の転換点とグレイン分布図の作成プロセス
抽象化した点に、各高さ別の集積密度の高さを与える

第二章　10のシミュレーションで都市を「見える化」する | 083

グレイン分布図を作成するプロセス

　まず、それぞれの都市の住宅地図上から、建築物の平面図の重心を求めて抽象化し、「粒子の点変換」を行う。次に各粒子を階数別の低層、中層、高層、超高層の4種にわけ、各点を図上にプロットする（Fig.4）。次にプロットした点の集積密度を3D化することで、"グレイン（密度によるまとまり）"の強さや"グレインの分布状態（まとまりの分布状態）"の様子を視覚化する。この方法を用いて生成された、それぞれの都市の一定範囲内における4層の起伏のついた面状図を「グレイン分布図」と呼ぶことにする（Fig.5）。

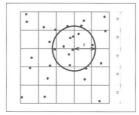

(Fig.4) グレイン分布図の領域
任意の点から半径30mの円内の
粒子密度を測定する

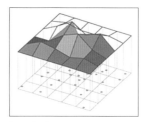

(Fig.5) グレイン分布図の例
15m間隔のグリッドの交点上に
密度の高さを表現する

　「グレイン分布図」は、都市の任意の点からある一定範囲（r）における分布した同質的粒子の密度を、その点の高さの係数として与えることによって、集積密度の強さをあらわし、これにより"さまざまな強さをもったグレイン"と"グレインの分布状態"を視覚化している。この一定範囲（r）は視力などの個人差を考慮に入れ、人が都市においてスケール的まとまりを知覚できる最小範囲である30mに設定した。（r=30m）また、ある一定の間隔（α）に分割した格子の交点において高さを与え、それらを繋ぐことで図示することにする。この格子の間隔（α）については、人間の下方視野角にしたがって、各都市共通に15mと設定した。

　また、東京、ロサンゼルスをはじめとして、現代の大都市といわれる街の規模は数十キロメートルにわたる都市域を形成しているが、「グレイン分布図」のカバーする大きさについては、槇文彦が述べている"半径数百メートルの都市像"を参考に、全ての都市共通の「グレイン分布図」の領域を600m×600mに設定した。最終的に、Auto Cad「AutoCAD」のプログラミング言語である「AutoLISP」Auto Lispを用いた「グレイン分布図作成システム」を構築することにより、コンピューターによる半自動的な算出方法で出力できるようにした（Fig.6）。次ページに、岡山県高梁市、兵庫県姫路市、東京都渋谷区円山町、東京都新宿区歌舞伎町を対象に「グレイン分布図」を作成した事例を示す（Fig.7,8,9,10）。

　以上、4つの地区について、シミュレーションを行ったが、高梁市と姫路は安定的なグレイン分布であるのに比べ、渋谷区円山町や、新宿区歌舞伎町では、混在化したグレイン分布であることが分かる。伝統的な都市と、近代になって飛躍的に開発が進行した都市のグレインの分布の違いがよく現われている。

　（Fig.11）は、都市の道路パターンの整合・不整合、グレイン分布の安定・不安定について、図式化したものであるが、計画された都市と、土着的な基底に近代的な開発が重層的に重なった都市では、グレイン分布の有り様が異なっていることが分かる。

	整合的	混合的	混沌的
都市組織			
グレイン分布図			

(Fig.11) 異なった都市ファブリックの街区群におけるグレイン分布図の現れ方
都市ファブリックの整合性とグレイン分布には密接な関係があり、
歴史的に重層的な地区の方がグレイン分布図の乱雑性が高い

　（Fig.12）はこれらの都市を2軸のマトリックスに示したダイアグラムであるが、高梁市や姫路市のように伝統的街並みが残っている都市のグレイン分布は安定しており、東京のように、多様な歴史の重層性のある都市ほどグレインの混とん性や不揃い性は高い。

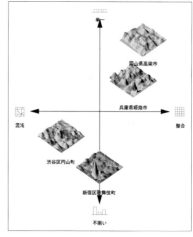

(Fig.12) グレイン分布図による
都市形態的類型を表したダイアグラム

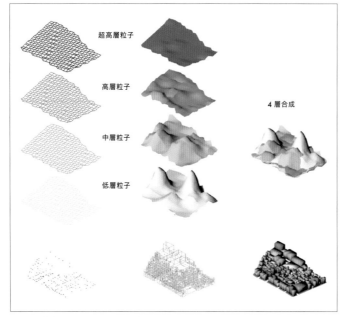

超高層粒子

高層粒子

中層粒子

低層粒子

4層合成

(Fig.6) 「グレイン分布図」の作成プロセス　各建物高さの分布図を個別に作成し合成する

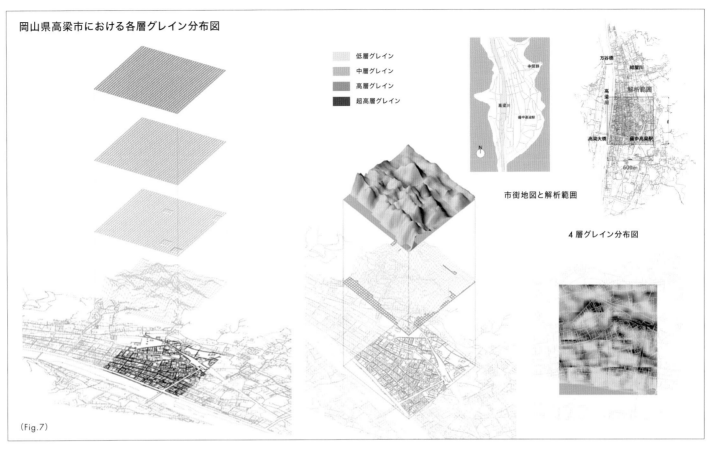

岡山県高梁市における各層グレイン分布図

低層グレイン
中層グレイン
高層グレイン
超高層グレイン

市街地図と解析範囲

4層グレイン分布図

（Fig.7）

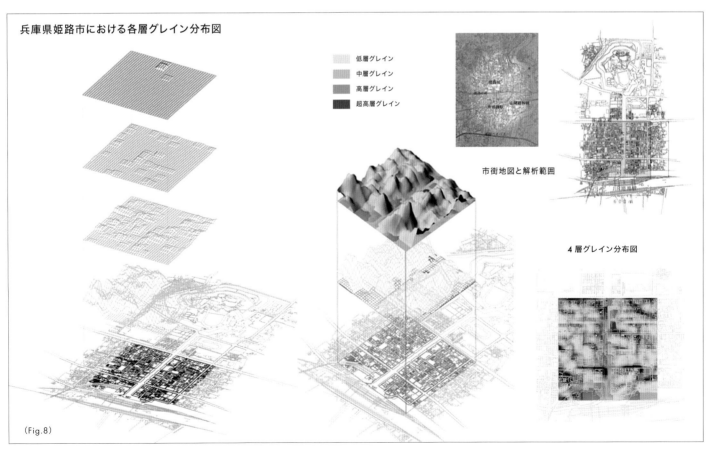

兵庫県姫路市における各層グレイン分布図

低層グレイン
中層グレイン
高層グレイン
超高層グレイン

市街地図と解析範囲

4層グレイン分布図

（Fig.8）

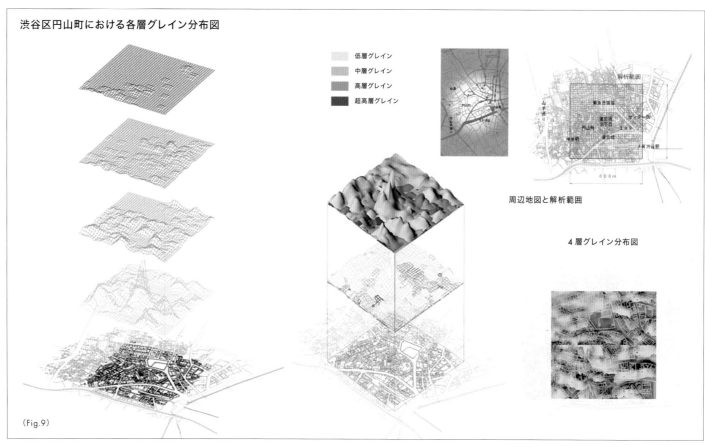

渋谷区円山町における各層グレイン分布図

低層グレイン
中層グレイン
高層グレイン
超高層グレイン

周辺地図と解析範囲

4層グレイン分布図

（Fig.9）

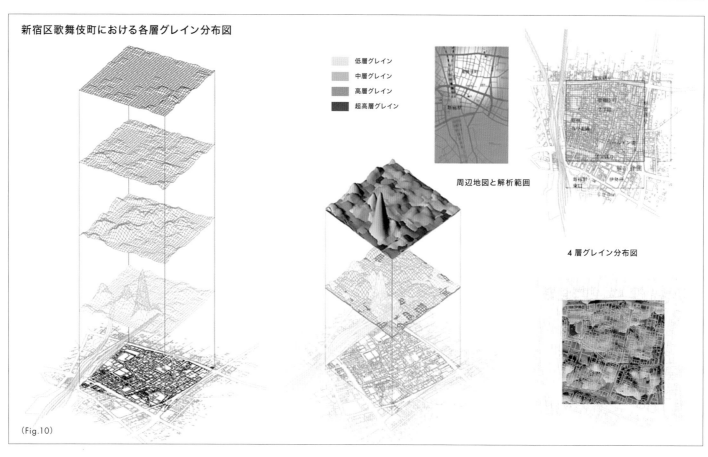

新宿区歌舞伎町における各層グレイン分布図

低層グレイン
中層グレイン
高層グレイン
超高層グレイン

周辺地図と解析範囲

4層グレイン分布図

（Fig.10）

Visualization of Urban Grains

SUMMARY

In this section, we further develop the density of accumulation graph we attempted in Takahashi City, Okayama Prefecture, focusing on the scale and density of the accumulation of architecture (particles) in urban space to show the morphological characteristics of that urban space. As an indicator, we introduce the theory of creating a "grain distribution map." To explore the effectiveness of this methodology, we attempt to visualize the urban structure of four characteristic cities (historical city - Takahashi City, Okayama Prefecture; regional core city - Himeji City, Hyogo Prefecture; dense city - Shibuya Ward and Shinjuku Ward, Tokyo). Using these visualized "grain distribution maps," we will discuss the structure of each city, its problems, and its individuality from a "cohesive" perspective.

The "grain" as an architectural assemblage

Settlements and cities are formed as collective spheres of life. Therefore, buildings do not exist as particles in isolation, but take on a collective form. In older settlements and cities, they were formed as "town blocks" surrounded by streets. However, since Japan's rapid postwar economic growth, there has been a rapid increase in the number of large buildings and skyscrapers. Cities have been unable to cope with the significant changes in the scale of these particles and have been transformed into a state different from that of the past, when the city had a certain orderly structure. In general, the morphological changes in street patterns are small, and the form of city blocks has not changed over the years, while the particles that are the constituent elements of city blocks have changed in function and scale in response to the passage of time and socioeconomic changes, creating a large gap between city blocks and particles.

The current urban planning frameworks, such as district plans and superblocks, may differ in scale, but their basic concepts are the same as those of the city block. What is perhaps more important is a planning framework and methodology that can consider "contemporary urban particles." Here, we consider the "grain" scale as one of the new planning frameworks that will take up the middle ground between the city and architecture. "Grain" refers to a collection of coherent urban particles in the modern age, or urban structure at an intermediate scale (mezzo-scale), about the size of a city block. In other words, a "grain" is an aggregation of coherent building particles of similar scale and density, and a city can be reconsidered as an aggregation of various "grains."

In the 1950s, American urbanist Kevin Lynch proposed the concept of "imageability," which asserts that the identity of a city can be recognized only when many people perceive a common "cohesion." This "cohesion" with "imageability" is a clear form of "grain," which is an aggregate of particles in a city, or a state of easy-to-understand distribution. In this section, we simulate the way particles accumulate in cities and consider the individuality of cities from the viewpoint of the "grain" concept.

Here, we attempt to visualize the urban structure by focusing on the scale and density of the accumulation of particles in the urban space. In this visualized diagram, the density of coherent particles distributed within a certain range from an arbitrary point in the city is given as a coefficient of the height of that point. With this height expression, the "grain," which is a grouping of coherent particles with a certain density, forms various types of

peaks (protrusions). The form of these peaks varies with the distribution state of the particles, and the height of the peaks is proportional to the magnitude of the density among the particles. Grains A and B in (Fig. 1) visualize two different densities of coherent particles, while Grains C, D, and E in (Fig. 2) visualize the same density of three different types of heterogeneous particles.

In this section, we introduce and expand this particle concept and attempt to visualize the increasingly complex modern urban structure by focusing on the scale and density of buildings (particles). We will consider the unique structure of each city by focusing on the "grain," which is an aggregate of buildings.

Process employed to create a grain distribution map

First, the center of gravity of the floor plan of the building is abstracted from the residential map of each city, and a "point transformation of particles" is performed. Then, each particle is divided into four types by number of floors (low-rise, medium-rise, high-rise, and ultra-high-rise), and each point is plotted on the diagram. Finally, the density of the plotted points is converted to 3D to visualize the strength and distribution state of the grains. The four-layered undulating areal map generated using this method within a certain area of each city is called a "grain distribution map."

The size of modern metropolitan cities such as Tokyo and Los Angeles form an urban area that spans several tens of kilometers, but the size covered by the grain distribution map was set to be the "image of a city with a radius of several hundred meters." As described by Fumihiko Maki, the area of the grain distribution map common to all cities was set to 600m x 600m with reference to the "Image of a City with a Radius of Several Hundred Meters." Finally, a "Grain Distribution Map Creation System" using "AutoLISP," the programming language of "AutoCAD," was constructed so that the data could be described in a graphical output using a semi-automatic computer-based calculation method.

The following are examples of "Grain Distribution Charts" created for Takahashi City, Okayama Prefecture; Himeji City, Hyogo Prefecture; Maruyama-cho, Shibuya-ku, Tokyo; and Kabukicho, Shinjuku-ku, Tokyo.

The simulation results for these four districts show that Takahashi City and Himeji have a stable grain distribution, while Maruyama-cho, Shibuya-ku, Tokyo and Kabukicho, Shinjuku-ku, Tokyo have a mixed grain distribution. The difference in grain distribution between traditional cities and those that have undergone dramatic development in the modern era is well illustrated (Fig. 10).

There is a close relationship between the consistency of the urban fabric and grain distribution, with historically more stratified neighborhoods having more cluttered grain distribution maps.

Fig. 11 presents these cities in a 2-axis matrix diagram. The grain distribution in cities with traditional townscapes, such as Takahashi and Himeji, is stable, while cities with diverse and layered histories, such as Tokyo, have a higher degree of grain crowding and unevenness.

Fig.1　Difference in height of grain distribution chart due to different densities
Fig.2　Image of overlap of grain distribution chart by different types of particles
Fig.3　Particle turning points and the process of creating a grain distribution chart
Fig.4　Region of grain distribution chart
Fig.5　Example of grain distribution chart
Fig.6　Grain distribution chart creation process
Fig.7,8,9,10　Grain distribution at 4 cities
Fig.11　Appearance of grain distribution maps for different city fabrics
Fig.12　Diagram of urban morphological typology with grain distribution

Chapter 2

8

公開空地の利用度と集積度
1. 公開空地の利用度を見える化する

概要

　国内の再開発制度のなかの総合設計制度（建築基準法）において東京都では、広場型、沿道型のような公開空地の類型に対し、異なった係数を与えているが、それが実際に有効であるのかをここでは検証している。公開空地を利用する人たちの行動集積を立体グラフで示し、実際の使われ方を分かりやすく表現する。

利用されていない公開空地の実状

　東京の建て詰まりで公共的なオープンスペースが恒常的に不足している状況を脱し、高層ビルの開発のボーナス制度と引き換えに公開されたオープンスペース（公開空地）を提供することを狙いとして総合設計制度が1970年に施行された。これは海外ではPOPS（Private Owned Public Space）と呼ばれているが、基本的に民有地が開放されているため、わが国では管理上の理由で、チェーンなどにより人の進入を排除するような違法行為もみられる。また、単に面積や接道長さなどの量的な仕様が認可時に評価されるため、ベンチや植栽などの空間の設えの質的な評価がされず、利用者の少ない公開空地も多くみられる。今後はニューヨークのPOPS制度のように、質に関する細かいガイドラインが計画の認可時に提示されるべきである。ここでは、東京都の沿道型や広場型のような公開空地の類型と係数制度を確認し、具体的な公開空地における利用度を視覚化することによって、それらの係数が妥当であるかを評価することを試みた。アメリカの活動家ジェー

ン・ジェイコブスが唱えたように、都市における街区はできるだけ小さく、人々が歩きまわりやすくするることが好ましい。最近はそのことをウォーカブルシティーと言ってまちづくりの中心的概念になってきた。ともするとスーパーブロックの再開発が多く展開されるなかで、ウォーカブルなまちづくりを推進していくためには、敷地内を通過する経路の確保を誘導することはますます重要になると思われる。そういう意味で、この公開空地の制度は道路のグリッドを外れたところに通貨動線が生まれるため、うまく利用されれば大変使い勝手の良い制度である。

　公開空地の種類には以下のカテゴリーがあり、国内の再開発制度において、これらの公開空地を併設することが基礎的なルールとなっている。

> **■ 総合設計制度（建築基準法）がつくる公開空地**
> 　計画建築物の敷地内の空地又は開放空間（アトリウム、ピロティなどおよび人工地盤等をいう。など）のうち、日常一般に公開される部分（中略）で別に定める公開空地の規模・形状の基準に適合する帯状または一段の形態をなすものをいう。（東京都総合設計許可要綱）
>
> **■ 特定街区制度（都市計画法）がつくる有効空地**
> 　『当該地区の環境整備に有効で公衆の使用に供する空地』。街区内の空地（公園、緑地、広場等）又は建築物の開放空間のうち、日常一般に開放された部分で、その面積が100平方メートルを超えているもの（東京都特定街区運用基準）
>
> **■ 高度利用地区制度がつくる広場等の有効な空地、**
> **　公共的屋内空間、壁面の位置の制限により確保する空地**
> 　「壁面の位置の制限により確保する空地」：『高度利用地区内に設ける壁面の位置の制限により確保する空地は、日常一般に公開される部分で、原則として歩道状に築造するもの』（東京都高度利用地区指定方針及び指定基準）

（Fig.1）

■ 一般に開放され、比較的利用されているケース

■ 管理されすぎて、あまり使用されていないケース

（Photo1）公開空地の利用度状況：民有地であるため、公開を渋る開発も多くみられる

現状を把握するため、いくつかの公開空地の調査を実施したが、利用されている空地、利用されていな空地の存在が明確に分かれていた（Photo1）。

次に、接道パターンと公開空地の平面パターンを類型化し、マトリックスによる分析する。ここでは、主に東京都港区の事例を示すが、基本的には、1）敷地と接道との関係条件、2）通り抜けなどの歩行者専用道の条件などが類型化される（Fig.2）。

a.1方向型	b. 分離 2方向型	c.L字 2方向型	d.3方向型	e.4方向型

1.貫通型	2.I型	3.二型	4.L型	5.凹型	6.口型	7.複合型

（Fig.2）公開空地の類型化：接道条件、敷地内の通り抜け条件などが示されている

公開空地有効係数

1.歩道状空地

歩道状空地（幅員が4m以下で道路との高低差が1.5m以下のものに限る）の有効係数は、計画する地域ごとに、連続（二辺以上の連続を含め、出入口などによる分断は必要と認められる範囲で連続とみなす）する歩道状空地の長さに応じて、下表の当該各欄に掲げる数値とする。その他の部分は：1.0（Fig.2）。

	L < 20	20≦L<40	40≦L<60	60≦L<80	80≦L<100	100≦L
都心部・副都心	1.5	1.7	1.9	2.1	2.3	2.5
環状7号線の内側	1.3	1.5	1.7	1.9	2.1	2.3
その他の区域	1.3	1.3	1.5	1.7	1.9	2.1

L：長さ（m）
都心部：区部中心部整備指針で定める「都心部」の区域
副都心：副都心整備計画及び臨海部副都心まちづくり推進計画に定める区域

（Fig.3）公開空地の有効係数
歩道状空地の係数が高くなっており、開発の条件としてインセンティブが与えられている

2.広場状空地（ピロティなどおよび人工地盤等の部分を除く）

ピロティ等及び人工地盤等の部分をのぞいた面積が300m² 以上の広場状空地で、幅員が6m以上の道路、歩道状空地又は貫通通路に接するもの
  a．道路等に面するもの：1.2
  b．道路等に面しない部分：0.6
以外のもの
  a．道路等に面するもの：1.0
  b．道路等に面しない部分：0.6

3.貫通通路

屋外貫通通路：1.0
屋内貫通通路：0.4 ～ 1.0（その規模、形態に応じて）

調査手順

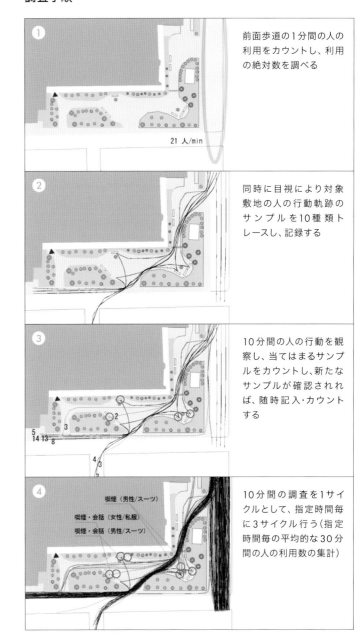

① 前面歩道の1分間の人の利用をカウントし、利用の絶対数を調べる

21 人/min

② 同時に目視により対象敷地の人の行動軌跡のサンプルを10種類トレースし、記録する

③ 10分間の人の行動を観察し、当てはまるサンプルをカウントし、新たなサンプルが確認されれば、随時記入・カウントする

④ 10分間の調査を1サイクルとして、指定時間毎に3サイクル行う（指定時間毎の平均的な30分間の人の利用数の集計）

喫煙（男性/スーツ）
喫煙・会話（女性/私服）
喫煙・会話（男性/スーツ）

（Fig.4）公開空地の有効係数
歩道状空地の係数が高くなっており、開発の条件としてインセンティブが与えられている

GIS を利用したグラフ化のプロセス

① 調査結果を反映した図面（2D）に適当な大きさのグリッド（1m）を書き込む

② 線データとなっている行動トレースラインを、グリッドで切り取る（GIS）

③ 切り取られたラインがグリッドの中にいくつ含まれているかをカウントする（GIS）

④ その数を高さに設定し、グリッドの中心同士を結んだ surface を立ち上げる（GIS）

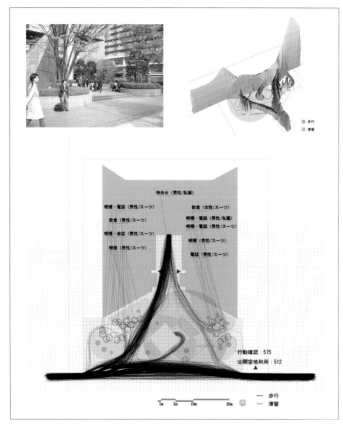

（Fig.5）日本新聞インキビル（東京都・品川駅前）のケース

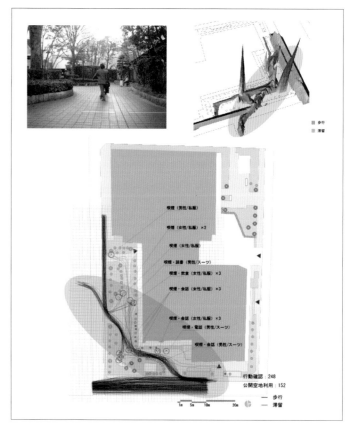

（Fig.6）芝超高層ビルのケース

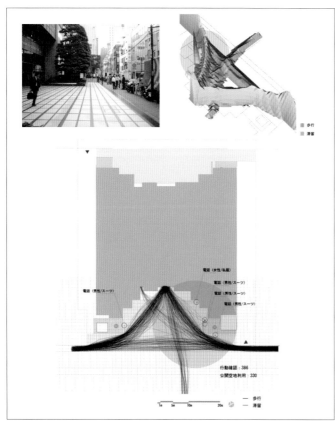

（Fig.7）日本橋プラザビル（東京駅付近）のケース

これらのシミュレーションから分かったことと提言

周辺環境

周辺環境には、「用途」「交通量」「歩行者数」などの変数があり、それらにより空地の利用度が大きく異なる。係数は一律でなく、エリアごとに設定するべきである。

ショートカットの有効性

「角地」…公開空地を角に配置することはノード（結び目）として重要であるので、係数割増しを定めることが望まれる。

協調型公開空地

協調して連担した公開空地を優遇し、係数割増しすることが考えられる。

歩道状空地

最も高く設定されている係数（1.3 〜 2.5）であるが、さらにベンチ長さによるインセンティブも考えられる。

貫通通路

一律で係数1.0であるが、ベンチを一定数設けることによる割増しを考えることが望まれる。

Visualization of the Degree of Use of Public Open Space

SUMMARY

The Tokyo Metropolitan Government, in its comprehensive design system (Building Standards Law) for redevelopment in Japan, assigns different coefficients to different types of open space, such as plaza-type and roadside-type spaces. This section examines whether this system is actually effective. A three-dimensional graph shows the accumulation of activities of people who use open space and represents the actual use of open space in an easy-to-understand manner.

The reality of underutilized public open space

The comprehensive design system was enacted in 1970 with the aim of providing public open space in exchange for a bonus system for the development of high-rise buildings in order to overcome the lack of public open space due to Tokyo's constant construction. In other countries, this is referred to as POPS (privately owned public space). Since private land is basically opened to the public under this system, some illegal acts have emerged in Japan in response to it, such as using chains to exclude people from entering for administrative reasons. In addition, since quantitative specifications such as area and length of access roads are evaluated at the time of approval, the qualitative evaluation of space provision such as benches and plantings is not done, and there are many open spaces with few users. In the future, detailed quality guidelines should be provided at the time of planning approval, similar to the POPS system in New York. Here, we attempted to evaluate the appropriateness of these coefficients by identifying types of open space and coefficient systems, such as Tokyo's roadside type and plaza type, and by visualizing the degree of use of specific open space. As the American activist Jane Jacobs advocated, urban blocks should be as small as possible so that people can walk around easily. Recently, this has become a central concept in urban development, known as the walkable city. In this sense, the open space system is a very user-friendly system if used well, as it creates a currency flow line off the street grid.

The following categories of open space are available, and the basic rules of the redevelopment system in Japan are to have these open spaces.

- Open space created by the comprehensive design system (Building Standards Law)
  Open space is defined as a vacant lot or open space (atrium, pilotis, and artificial ground) within the site of a planned building that is open to the public on a daily basis (omitted), and that is in the form of a strip or a series of strips that conform to the scale and shape standards for open space specified separately. (Tokyo Metropolitan Government Comprehensive Design Permit Outline)

- Effective open space created by the specific district system (City Planning Law)
  Vacant land that is used by the public and is effective in improving the environment of the area in question. Vacant land (parks, green spaces, and plazas) or open spaces of buildings within a city block that are open to the public on a daily basis and whose area exceeds 100 square meters (Tokyo Metropolitan Government's Specific City Block Operation Standards)

- Effective open spaces such as plazas, public indoor spaces, and open spaces secured by the restriction on the position of walls, which are created by the high-use district system. Vacant land to be secured by restricting the location of walls: "Vacant land to be secured by restricting the location of walls in high-development use districts shall be open to the public on a daily basis and, in principle, shall be constructed in the form of a walkway."

(Tokyo Metropolitan Government High-Level Use District Designation Policy and Criteria))

In order to understand the current situation, we conducted a survey of several open spaces and found that there was a clear distinction between utilized and unutilized open spaces.

Next, a matrix analysis was conducted by categorizing road patterns and the patterns of open space plans. Fig. 1 presents a case study of Minato Ward, Tokyo, which is basically categorized by 1) the relationship between the site and the street, and 2) the condition of pedestrian paths such as through streets.

Open Space Effectiveness Factor

1. Sidewalk open space

The effective coefficient of open space with sidewalks (limited to those with a width of 4 m or less and an elevation difference from the road of 1.5 m or less) shall be the values listed in the respective columns of the table (Fig. 2) for each planned area, depending on the length of the open space with sidewalks that are continuous (including those with continuous areas on two or more sides; separation by an entrance or exit shall be considered as continuous to the extent deemed necessary). Other areas: 1.0

2. Plaza-type open space

Square-shaped vacant land with an area of 300 m2 or more, (excluding pilotis and artificial ground) and with a width of 6 m or more, a pedestrian walkway, or a passage.
  a. Facing a road: 1.2
  b. Areas not facing a road: 0.6
other than above
  a. Facing a road: 1.0
  b. Areas not facing a road: 0.6

3. Walk-through passageway

Outdoor passageways: 1.0
Indoor walkways: 0.4 to 1.0 (depending on the scale and form of the walkway)

Below are findings and recommendations from these simulations.

- Surrounding environment
  The surrounding environment includes variables such as "use," "traffic volume," and "number of pedestrians," and the degree of use of vacant land varies greatly depending on these variables. The coefficient should be set for each area, not in a uniform manner.

- Effectiveness of shortcuts
  Corner sites: It is important to place public open space at the corner as a node (knot), so a coefficient premium should be determined.

- Cooperative open space
  Preferential treatment is given to public open spaces that are linked in a cooperative manner, and the coefficient is increased.

- Sidewalk open space
  The highest coefficient is set (1.3 to 2.5), but further incentives based on the length of benches could also be considered.

- Walk-through passageway
  The coefficient is uniformly set at 1.0, but it is desirable to consider an increase in the coefficient by providing a certain number of benches.

Photo1 The degree of use of publicly available land
Fig.1   Typology of open space, including conditions for access and pedestrian walk-throughs.
Fig.2   Effective coefficient of open space
         The coefficient for sidewalk-like open space is higher and is incentivized as a condition for development
Fig.3   Effective coefficient of open space
         The coefficient for sidewalk-like open space is higher and is incentivized as a condition for development
Fig.4   Case of Nihon Shinbun Ink Building (in front of Shinagawa Station, Tokyo)
Fig.5   Case of super high-rise building at Shiba
Fig.6   Case of Nihonbashi Plaza Building (near Tokyo Station, Tokyo)

Chapter 2	公開空地の利用度と集積度
8	**2. オープンスペースの分布を見える化する**

概要

総合設計制度および特定都市計画制度により、都内の公開空地が確実に増え続け、都市生活者に貢献していることは好ましいが、超高層ビルの建設とトレードオフで開発され、使い勝手が悪いといわれる空地の「質」の問題はこれから議論を続けるべき大きな課題である。19世紀に、アメリカの造園家・都市計画家のフレデリック・ロー・オルムステッド（1822-1903年）が米国ボストンで手掛けたように、これらが緑空間の連鎖として二次元的にネットワーク化され、生態学的連鎖が生まれるようなビジョンはまだ描かれていない。ここでは、都内各区における公開空地の位置関係を把握し、それらの集積度を視覚化する。

公開空地の連続性

現在、都市計画における公共のオープンスペース（公開空地）の整備計画や、民間による再開発事業に付随する公開空地の整備計画が個別に立案されており、総合的にコントロールされていないため、場当たり的な整備であることは否めない。今後の都市環境を考えるにあたり、それらを横串にした総合的なビジョンが必要であることは言うまでもない。現存の緑地や公園などの分布に、公開空地や有効空地の分布を重ねて、全体の将来ビジョンを視覚化することの意味は大きい。ここでは、生態系も考慮し、緑地のつながりを強化することで、点であったオープンスペースをネットワーク化して線や面として機能させ、今後の都市環境に有効に利用されていく方法を考える。

ここでは、具体的にGISを用いて、さまざまなデータのプラットフォームを構築し、空地や緑の集積度などを視覚化する。

データ の種類　　項目	作成方法	基となるデータ
公開空地 分布図	「東京都総合設計許可要綱とその解説 改訂4版」の公開空地の住所を基に、現地調査により把握した。千代田区、中央区、港区の公開空地をGIS上に記入	現地調査
有効空地 分布図	千代田区、中央区、港区の特定街区の有効空地を、現地調査と、東京都作成の「特定街区事例集」を基に把握したものを、GIS上に記入	現地調査 「特定街区事例集」
緑地分布図	東京都の平成9年度国勢調査による、土地利用の公園や、政府所有の緑地や、スポーツ施設の公園など をピックアップしGIS上に記入	東京都の平成9年度国勢調査に現地調査を加えたもの
建物床面積	東京都の平成9年度国勢調査による建物面積と階数 を乗じたもの	東京都の平成9年度国勢調査に現地調査を加えたもの
鉄道駅	東京都区分地図から駅の位置を把握し、GIS上に記入	東京都区分地図昭文社

（Fig.1）

公開空地分布図作成のためのデータ

都心3区である千代田区、中央区、港区の公開空地について、「東京都総合設計許可要綱とその解説 改訂4版」の公開空地の住所を基に、実際に現地に赴き公開空地を調査した。掲示板などで確認を行った「公開空地」の地図を基に、GIS上にそれぞれの公開空地の形状を記入している（Fig.1,2）。

分析の手法

これらのデータを使用してGISを用いて分析を行った。分析方法は、大きく分けると以下の3つである。

1）ベクターモデルによる分析（地図データ）
2）メッシュデータによる分析
3）3Dデータ（TINモデル）による分析

これらは分析の目的によって使い分けて使用することで、より深く考察を行うことが可能である。特にネットワークという概念を用いるために、3Dデータを重要視している。

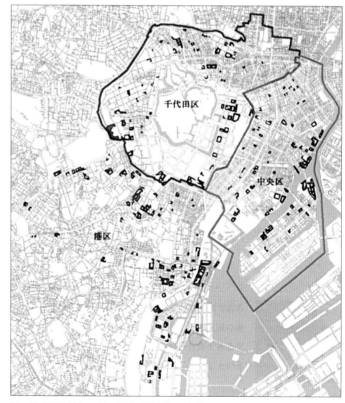

（Fig.2）調査結果をマッピングした公開空地の分布図

1）ベクターモデルによる分析（地図データ）

いわゆるベクターデータであり、GIS上に、建物形状、歩道形状、公開空地形状などが記入され、それぞれの細かい形状が把握できる。街区単位など細かいスケールの分析や、歩道の関係など厳密性を要する場合に使用する（Fig.2）。

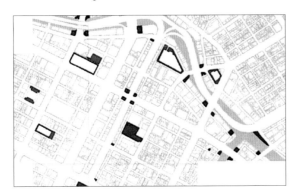

（Fig.3）ベクターデータによる公開空地の分布図

2）メッシュデータによる分析

単純で規則的な矩形状の地域単位で地表面を仕切り、その区画ごとにデータを集計したものを、通常メッシュデータと呼ぶ。メッシュデータは広域的な分析に適しており、ここではセルと呼ばれる規則的に並んだ四角形に属性をもたせるモデルを用いる。そして、ひとつのセルの中に要素がどれくらい集積しているかによって属性を表したものを使用し、2D上や3D上で表示する（Fig.3）。

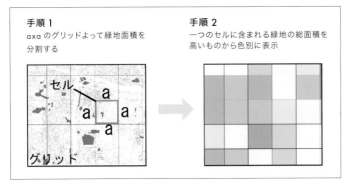

（Fig.4）メッシュデータによる緑地の分布図

3）3Dデータ（TINモデル）による分析

上記のメッシュデータのセルの値を高さに反映し、それらを等高線のようにつなげて表示する方法である。等高線のZ方向の傾きから隣り合うセルの差を視覚化でき、それを位置情報として表示することができる。メッシュデータがある地域を分析するのに適しているのに対して、TINデータでは地域間の関係性を把握する際に最も適したデータである。特につながりや、まとまりについて考察を行う際に有効である（Fig.4）。

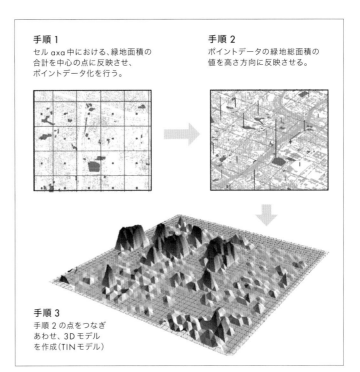

手順1
セルaxa中における、緑地面積の合計を中心の点に反映させ、ポイントデータ化を行う。

手順2
ポイントデータの緑地総面積の値を高さ方向に反映させる。

手順3
手順2の点をつなぎあわせ、3Dモデルを作成（TINモデル）

（Fig.5）TINモデルによる緑地の分布図

まずは、都心部の住宅の分布状況を調べてみる。
最終的には、TINモデルで高さ方向を与えることで、誰にも分かるグラフにすることができる（Fig.6）。

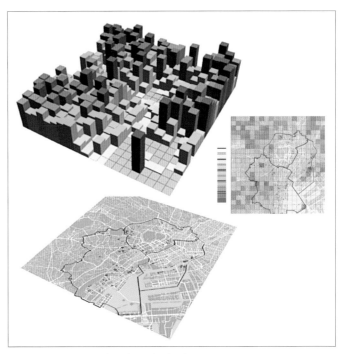

（Fig.6）分析方法の検討例（都心部の住宅分布図）
それぞれの目的に応じてグリッドや分析方法の比較検討を行なうことができる

　以下は、まず東京都の千代田区、中央区、港区の既存の「緑地および公園」の地図に「公開空地および有効空地」の分布を合成したものである（Fig. 7, 8, 9, 10, 11, 12）。

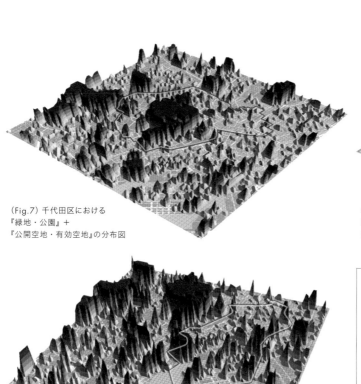

（Fig.7）千代田区における
『緑地・公園』＋
『公開空地・有効空地』の分布図

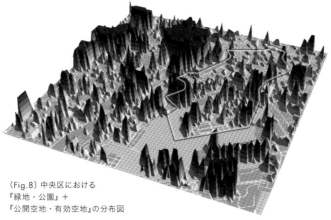

（Fig.8）中央区における
『緑地・公園』＋
『公開空地・有効空地』の分布図

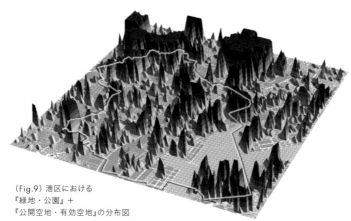

（Fig.9）港区における
『緑地・公園』＋
『公開空地・有効空地』の分布図

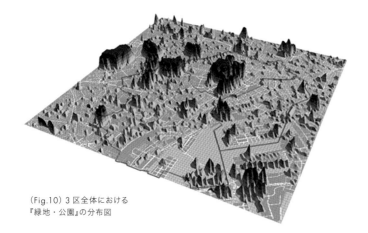

（Fig.10）3区全体における
『緑地・公園』の分布図

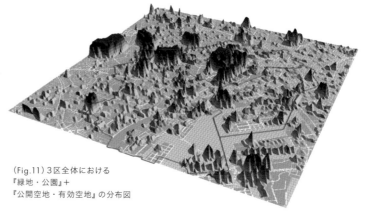

（Fig.11）3区全体における
『緑地・公園』＋
『公開空地・有効空地』の分布図

各要素の影響力：重ね合わせた図

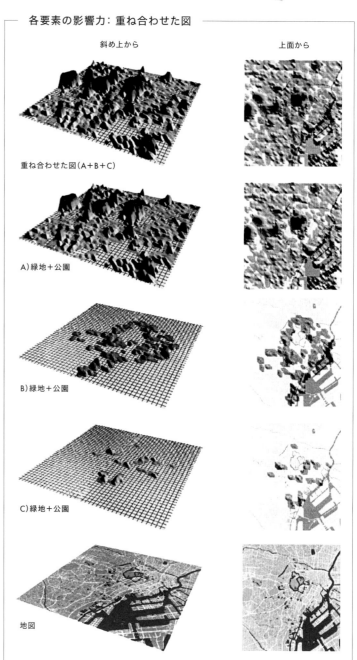

斜め上から　　　　　　　　　上面から

重ね合わせた図（A＋B＋C）

A）緑地＋公園

B）緑地＋公園

C）緑地＋公園

地図

（Fig.12）

Visualization of the Distribution of Public Open Space

SUMMARY

Although it is favorable that the number of public open spaces in Tokyo is steadily increasing and contributing to city dwellers' wellbeing through the comprehensive design system (Sogo Sekkei seido) and the specific city planning system (Tokutei Gaiku seiido), the issue of the "quality" of open space, which is developed in trade-off with the construction of skyscrapers, is said to be less usable. In this section, we identify the location of public open space in each ward of Tokyo and visualize the degree of concentration of these spaces.

Continuity of public open space

Currently, plans for the development of public open space in urban planning and plans for the development of open space associated with redevelopment projects by the private sector are drawn up separately and are not comprehensively controlled. Needless to say, there is a need for a comprehensive vision that transversely links all of these elements when considering the future urban environment. It is highly significant to visualize an overall vision for the future by overlaying the distribution of open space by the comprehensive design system and effective open space by the specific city planning system on the distribution of existing green spaces and parks. In this section, we also consider ecosystems and how to strengthen the connection of green spaces so that open spaces that were once points can be constructed into networks and function as lines and planes, and be used effectively in the urban environment in the future. We will specifically use GIS to build a platform for various data and visualize the concentration of open spaces and greenery.

Data for creating a distribution map of publicly available vacant land

We surveyed open space in the three central wards of Tokyo, Chiyoda, Chuo, and Minato, based on the addresses of open space in the "Tokyo Metropolitan Government Comprehensive Design Permission Manual and its Commentary, Revision 4," and visited the sites to survey the open spaces. The shape of each open space was input into GIS.

(1) Distribution Map of Specific City Block

Similar to the open space distribution map, the shapes of the specific city block open spaces were input in the GIS based on out site visits. We also made assumptions based on the "Collection of Examples of Specified City Blocks" prepared by the Tokyo Metropolitan Government.

(2) Green space distribution map

A green space distribution map was created by choosing parks of land use according to the Tokyo Metropolitan Government's census, government-owned green spaces, and parks of sports facilities.

Analysis Method

Using these data, GIS was used to conduct the analysis. The analysis methods can be broadly classified into the following three categories.

(1) Analysis by Vector Model (Map Data)

This is so-called vector data, in which the shapes of buildings, sidewalks, and public open space are entered in GIS, and the detailed shape of each can be determined.

(2) Analysis by mesh data

Mesh data are simple, regular, rectangular units that divide the earth's surface into sections and aggregate the data for each section. Mesh data are suitable for wide-area analysis. Here, we use a model in which attributes are attached to regularly arranged rectangles called cells. The attributes are expressed in terms of the number of elements clustered in a cell, and are displayed in 2D or 3D.

(3) Analysis using 3D data (TIN model)

This method reflects the values of the above mesh data cells in terms of height and displays them so that they are connected like contour lines. The difference between adjacent cells can be visualized from the slope of the contour lines in the Z direction, which can be displayed as location information. While mesh data are suitable for analyzing a certain region, TIN data are most suitable for understanding the relationships among regions. This is especially useful when considering connections and cohesion.

First, we will examine the distribution of housing in the city center.
Finally, by using the TIN model to give the height direction, we can create an easily understandable graph.

The following is a map of existing "green spaces and parks" in the Chiyoda, Chuo, and Minato wards of Tokyo, combined with the distribution of "open and available space" (Fig. 6).

Fig. 1　Distribution map of open space mapping the survey results
Fig. 2　Distribution of public open space using vector data
Fig.3　Distribution map of green areas using mesh data
Fig.4　Distribution of green spaces based on the TIN model
Fig.5　Example of examining the analysis method (distribution map of houses in the city center)
Fig.6　Distribution of "green spaces and parks" + "open space and available space" in Chiyoda Ward
Fig.7　Distribution of "green spaces/parks" + "open space/available space" in Chuo Ward
Fig.8　Distribution of "green spaces and parks" + "open space and effective open space" in Minato-ku
Fig.9　Distribution of "green spaces and parks" in all three wards
Fig.10　Distribution of "green spaces and parks" + "open space and effective open space" in the three wards

Chapter 2

9

東京都における超高層開発の分布予測
超高層開発の分布を見える化する

概要

　東京都の総合設計制度および特定街区制度では、公開空地とセットで開発することで、建築基準法で規制されている高さや容積制限が緩和される。また、90年代の小泉政権移行「特区制度」が施行され、都内では大型開発の動きが止まらない。ここでは、これらの制度が可能と思われる大規模敷地を抽出し、今後の開発傾向の予測と分布予測を行う。最終的なインプットデータは若干古いが、約20年前の2003年にシミュレーションした予測がほぼ現状を指し示し、分析予測が有効であったことが確認された。

ボストンの都市計画

　都市の景観については、アメリカのボストンでは先進的な取り組みをしてきたことで有名である。ボストンは、独自の都市再開発政策を導いてきたため、都市内に豊富な歴史的・環境的資源を保全したまま、現代的な都市の姿へと成長してきた（Photo1）。その成長過程の中で最も重要視されてきたものはコミュニティ形成、つまり都市の中で暮らす市民の政策参加である。ボストンの都市計画では、バランスのとれた「成長」を導こうという基本政策に基づき、開発による都市の成長と共に、質の高い都市のデザインの追及が行われている。そして、その政策は、経済の成長を管理するとともに、住宅や雇用などの社会問題も同時に改善させてきたことが特徴的である。その背景に流れているのは、コミュニティ住民の利益、都市の成長のもたらす都市生活全体への影響の配慮、およびコミュニティと一緒に策定されたプランが最も

望ましい結果をもたらすという共通の意識であった。
　ボストンの都市景観を決定する上で非常に重要な要素にゾーニングがある。ボストンはコミュニティを中心とした小さなまとまりである近隣住区という概念で都市が考えられてきた。そして、それらを包括するガイドラインの役割を担っているのが、「ダウンタウンゾーニング」計画と呼ばれるものである。その特徴として挙げられるのは、雇用やマイノリティーなどの社会的要素も都市計画の中に取り入れていること、道路幅員や斜線規制などとは独立して、地域ごとに許容絶対高さを定め、都市の物理的形態をコントロールする権限を持たせていることである。特に、後者の都市の形態に関する規制は、ボストンの特徴でもある低層の街並みといった伝統的都市環境を保全するために重要な役割を果たした。伝統的都市の保全のために開発が可能となる地域を制限することにより、逆に開発が可能な都心部における大型開発が推進されたのである。
　『都市のイメージ』(The Image of the City)という独自の都市論を示したことで知られているアメリカの都市計画家のケビン・リンチ（1918〜1984年）はボストンの都市計画に対して大きな影響を与えた。(Fig.1)は、リンチが提案したボストンの「成長のイメージ」で、ゾーニングにより都市形態をコントロールする事を示した。この政策により、都市のいたるところに超高層建築が無秩序に林立してしまった東京とは異なり、ボストンの街は形態にまとまりをもった、都市景観が創造されていることがよくわかる。
　都市において再開発が行われてきた背景としては、1）既成市街地における老朽化した木造密集市街地などの防災上の問題を多く抱えている地域の空地や道路の整備改善、2）欧米諸国の都市に比べて極めて少ない緑地空間の創出、3）少子高齢化社会を迎え、老人や子どもが安

（Photo1）ボストンの超高層群

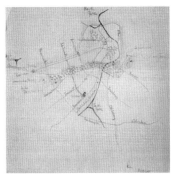

(Fig.1) ケビン・リンチのスケッチ

心して暮らせるバリアフリーを取り入れたまちづくりや公共施設の補完、4)良好な建築や住宅の整備、5)土地の有効かつ高度利用の推進、などが挙げられる。これまでも数多くの再開発事業が、都市環境向上を目的として行われてきたが、既成市街地におけるさまざまな問題点は未だ十分に解決されておらず、都市再開発はこれからも大きな役割を担うと考えられる。

しかし、これまでの再開発（特に大型再開発）は、それぞれの開発計画ごとの関係性が希薄であり、制度も個別の建築の性能評価のみの規定しかないために、その結果、高層ビルが林立し、都市景観を荒廃させてしまったように思われる。また、これからも社会的な要求から超高層建築が継続して建設されると予想されるため、その計画に都市全体を配慮した何らかの緩やかなコントロールを加える必要があるのではないのだろうか。

再開発と超高層を語るうえで無視できない要素の1つが、その形態規制の緩和を可能としている法制度である。1963年の建築基準法改正による容積率制の導入に始まり、特定街区制度、総合設計制度、地区計画など、数々の法改正と規制緩和が繰り返されることで、都市の超高層化は進められてきた。そういった法制度の着目すべき点は、基本的に敷地に公共施設や空地の確保によって、形態や容積の緩和が受けられるようにするということである。

そこで、ここでは、再開発プロジェクトにおける重要な要素である超高層建築に着目し、超高層建築の都市景観に対する影響を、3Dモデルを用いて都市形態の継時的変化を基に考察する。また、そこまでの分析と考察を踏まえ、3Dモデルを用いて、都心3区の将来像に対するシミュレーションを行う。その条件としては、1)まず現状に対しての将来像を表現することを試み、2)その後、異なったシナリオによる3つのパターンの提案を行う。将来像のシミュレーションとしては、敷地の場所を都区内から求め、近年5,000m²以上の建替えや開発の行われていない敷地を選別し、そこへ容積率を高く設定した建物のモデルを作成し、このまま開発が行われた場合を想定した将来像を提示する。

基礎データについて

ここで作成した3Dモデルは「ゼンリン住宅地図」などを基に、5年おきに1971年から2001年まで、そして現況として2003年の地図データをGISへ入力し、それに高さを与えて作成した。また、特定街区及び総合設計制度適用建築物の情報は、「東京都総合設計許可要綱とその解説 改訂4版」「東京都特定街区運用基準」「建築統計年報」などから、住所や建物高さ、階数といった情報を得て作成した。データ作成の際、それらの資料に掲載されていた建物の名称が現在のものと異なっていたり、地番が変更されて見つけることができなかったりしたものがいくつかあるが、ほぼ全ての総合設計及び特定街区制度適用建物を網羅できた。また、全ての建物に関して、正確な建物高さの情報の把握は不可能であったため、住宅地図から読み取れる建物階数に、階高を4mとし、それを掛け合わせることによって得られた数値を建物高さとした。

3Dモデルは、地形・水部・緑地、広場・建物のレイヤで構成されている。そのうち建物は各年度3つのレイヤ（変化のしなかった建物・新しく建った建物・壊された建物）から構成されており、レイヤの表示を切り替えることで都市形態の変化を表現している。以下は、異なったシナリオによるシミュレーションである（Fig.2,3,4,5）。

シミュレーション1：無計画案

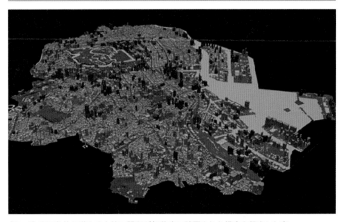

（Fig.2）単に条件を満たす土地に開発が無秩序に計画された場合を想定した案

シミュレーション2：都心部と尾根部

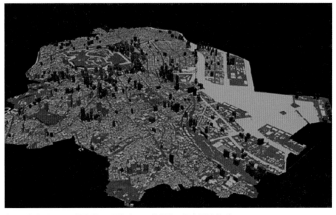

（Fig.3）皇居周りの都心部と、尾根（リッジ）部分に超高層建物が集中的に計画されることを想定した計画

シミュレーション3：尾根部集中案

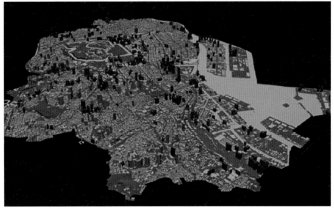

（Fig.4）尾根（リッジ）部分だけに超高層建物が集中的に計画されることを想定した計画

シミュレーション 4：都心部およびウォーターフロント案

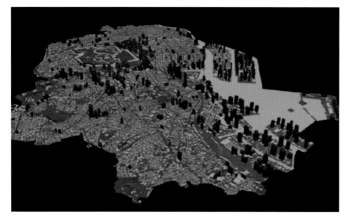

（Fig.5）皇居周りの都心部と、湾岸沿いの敷地に集中的に計画されることを想定している
※赤＝既存　※5000m²以上の敷地で、1971年以降建替えが行われなかった敷地に超高層が建設された場合

CHAPTER 2, 9 — PREDICTING THE DISTRIBUTION OF HIGH-RISE DEVELOPMENT IN TOKYO

Visualization of the Distribution of Super High-rise Development

SUMMARY

Under the Tokyo Metropolitan Government's Comprehensive Design System and Specified City Block System, height and volume restrictions regulated by the Building Standards Law has eased for development in combination with public open space. In addition, with the implementation of the "special zone system" under the transition of the Koizumi administration in the 1990s, there has been an unstoppable trend toward large-scale development in Tokyo. In this section, we identify large-scale sites in which these systems are possible and forecast future development trends and distribution. Although the final input data is slightly out of date, the forecast simulated in 2003, about 20 years ago, almost pointed to the current situation, confirming that the analytical forecast was valid..

Urban Planning in Boston

With regard to urban landscapes, Boston, USA is well known for its progressive approach. Boston has led a unique urban redevelopment policy that has allowed the city to grow into a modern urban form while preserving its abundant historical and environmental resources. One of the most important aspects of this growth process has been community building, or policies encouraging the participation of citizens living within the city. Based on the basic policy of guiding balanced "growth," Boston's urban planning has pursued high-quality urban design along with urban growth through development. Its policies are

　（Fig.6）は2023年の段階のウォーターフロントの超高層ビルの分布図である。ほぼ、都心部およびウォーターフロントに集中して建てられており、「都心部およびウォーターフロント」のシミュレーションが予測した通りに開発が進んだことが分かる。これは、港区などの尾根部分には、既存住民が多く居住しており、超高層ビルが建てられにくい状況にあること、逆に千代田区の皇居周りや江東区の湾岸地域には既存住民が少なく、超高層ビルが比較的建てやすくなっていることが挙げられる。

　今後、超高層ビルの乱立を防ぐためには、こうした知見を生かし、ケビン・リンチがアメリカ・ボストンで実施したように、単に乱立することを許さないメリハリのある都市景観政策を検討していくことが望まれる。

（Fig.6）現状の超高層ビルの分布図

characterized by their ability to manage economic growth while simultaneously improving social issues such as housing and employment. Underlying this has been a shared awareness of the interests of community residents, a concern for the overall impact of urban growth on urban life, and a belief that plans developed with the community produce the most desirable results.

Zoning is a very important factor in determining Boston's urban landscape. Boston has been conceived of as a city of neighborhoods, which are small, community-centered groups. The "Downtown Zoning" plan, as it is called, serves as an overarching guideline for these neighborhoods. The characteristics of these plans include the incorporation of social factors such as employment and minorities into urban planning and the establishment of permissible absolute heights for each area, independent of street width and diagonal regulations, giving the city the authority to control its physical form. In particular, the latter regulation on urban form has played an important role in preserving the traditional urban environment, such as the low-rise streetscapes that characterize Boston. By restricting the areas that could be developed in order to preserve the traditional urban environment, the city has been able to promote large-scale development in the urban centers where development was possible.

MIT Professor Kevin Lynch, known for his original urbanism, "The Image of the City," has had a major influence on Boston's urban planning. Fig.1 presents Lynch's proposal for an "image of growth" for Boston, which would control urban form through zoning. This policy has created a cohesive urban landscape in Boston, unlike Tokyo, where skyscrapers are randomly popping up in every corner of the city.

Trends and Simulations of Urban Redevelopment in Urban Centers

The background of urban redevelopment includes: 1) improvement of vacant lots and roads in existing urban areas that have many disaster prevention problems, such as aging, dense wooden urban areas; 2) creation of green spaces, which are extremely scarce compared to cities in Europe and the United States; 3) development of barrier-free communities and supplementary public facilities in which the elderly and children can live safely in an aging society with a declining birthrate; 4) development of good buildings and housing; and 5) promotion of effective and high-level use of land. Although numerous redevelopment projects have been undertaken to improve the urban environment, various problems in existing urban areas have yet to be fully resolved, and urban redevelopment is expected to continue to play a major role in the future.

However, redevelopment projects (especially large-scale redevelopment projects) to date have not been well connected to each other, and the system only provides for performance evaluation of individual buildings, resulting in a forest of high-rise buildings that have devastated the urban landscape. Moreover, since it is expected that super high-rise buildings will continue to be constructed in the future due to social demands, it may be necessary to add some kind of gradual control to their planning that takes the city as a whole into consideration.

One factor that cannot be ignored when discussing redevelopment and super high-rise buildings is the legal system that allows for the relaxation of restrictions on architectural form. The main point to note about these legal systems is that they are based on basic principles: their focus is to ensure that public facilities and vacant land are secured on the site so that the form and volume of buildings can be eased.

In this section, we focus on high-rise buildings, which are an important element in redevelopment projects, and examine their impact on the urban landscape, based on the changes in urban form over time using a 3D model. Based on the analysis and discussion up to this point, a simulation of the future image of the three central wards of Tokyo will be conducted using the 3D model. The conditions for this simulation are: 1) first, an attempt to express the future image of the current situation, and 2) then, three different scenarios will be proposed. The simulation of the future image is based on a selection of sites in the Tokyo metropolitan area that have not been reconstructed or developed in recent years and are 5,000 m2 or larger, and a building model with a high floor-area ratio is created to present a future image assuming that development continues as is.

Basic Data

The 3D model created here is based on "Zenrin Residential Maps." Map data from 1971 to 2001 at five-year intervals and the current status in 2003 were input into GIS, as was the height of the model. Information on specific districts and buildings subject to the comprehensive design system was obtained from the "Tokyo Metropolitan Government Comprehensive Design Permit Outline and its Commentary, Revised 4th Edition," "Tokyo Metropolitan Government Specific District Operation Standards," and "Annual Report of Building Statistics" for addresses, building height, and number of floors. Although the names of some of the buildings listed in these documents were different from the current names, and some of the lot numbers were changed and could not be found, we were able to cover almost all buildings that were subject to the comprehensive design and specific district system. The 3D model consists of layers of topography, water areas, green areas, plazas, and buildings. The buildings consist of three layers for each year (unchanged buildings, newly constructed buildings, and destroyed buildings), and the changes in urban form are represented by switching the display of the layers.

The following are simulations of different scenarios:

(Fig.2) No-plan scenario
A scenario in which development is simply planned in an unregulated manner on land that meets the conditions.

(Fig.3) City center and ridge
A plan that assumes that high-rise buildings are planned to be concentrated in the city center around the Imperial Palace and in the ridge area.

(Fig.4) Concentration on the ridge
A plan that assumes that high-rise buildings will be concentrated only on the ridge portion.

(Fig.5) City Center and Waterfront Plan
A plan that assumes that high-rise buildings will be concentrated in the city center around the Imperial Palace and on sites along the waterfront.

(Fig.6) Distribution of current skyscrapers (Google Map)

Almost all of the buildings are concentrated in the central Tokyo area and along the waterfront, indicating that development has progressed as predicted by the "central Tokyo area and waterfront" simulation. This is because many residents live on the ridges in Minato Ward and other areas, making it difficult to build skyscrapers, while there are few existing residents around the Imperial Palace in Chiyoda Ward and in the bay area in Koto Ward, making it relatively easy to build skyscrapers.

In order to prevent the proliferation of skyscrapers in the future, it is desirable to utilize these findings and consider a well-defined urban landscape policy that does not simply allow them to grow wildly, as American urban planner Kevin Lynch (1918-1984) did in Boston.

Fig.1   Sketch by Kevin Lynch

Photo 1 Skyscrapers in Boston

Chapter 2 / 10 市民参加とエリアマネジメント
エリアマネジメントの育成プロセス

特別寄稿　泉山塁威

兵庫県姫路市の姫路駅前広場および周辺地域の整備は、ワークショップなどの市民参加型プロセスで進めることができた（詳細は三章124頁）。この空間におけるエリアマネジメント育成プロセスについて俯瞰して整理する。2014年当時の姫路市では、エリアマネジメントが成熟したわけではなく、まだなお発展途上で、地元行政や関係主体の尽力のもと、試行錯誤が続いている状況であった。駅前広場はそれだけでエリアマネジメントが完結するわけではなく、駅前はまちなかの玄関であり、ひとつの核にすぎないし、これから駅前とまちなかを双方考えたエリアマネジメントがスタートしようという段階であった。姫路駅前広場のハード整備に関するプロセス、およびそのプロセスが市民公開により進められてきたプロセスと併せて、どのようにエリアマネジメントや公共空間の活用の議論や仕組みがつくられたかを記述する。

ここでは、エリアマネジメントの育成プロセスの整理の方法として、2つの視覚化手法を取り上げる。1つ目は、これまで主に議論されて来たことや広場活用などの社会実験を列挙し、主体や属性別にカテゴライズし、心電図のようなポリライン（折れ線）で視覚化し、傾向を把握する手法である（Fig.1）。2つ目は、主体別の活動量をインフォグラフィックス（infographics）で視覚化し、年月の感覚を均一にしたなかで、月毎の活動量とその密度を視覚化した手法である（Fig.2）。

両図とも共通して、左から1）姫路市公式の推進会議、2）姫路市主催ワークショップ、3）商店街連合会主催勉強会・ワークショップ、4）地元NPO法人スローソサエティ協会主催セミナー、5）ワークショップ、6）勉強会、7）連絡会や協議会などの会議体、8）市民公開のフォーラム、9）公共空間活用の社会実験、10）公共空間の本設活用（マネジメント開始）というカテゴリーとしており、図の左から見ると、市の公式的な動きから、商店街、NPO、多主体の集まる会議体、市民公開の

フォーラム、社会実験と、図の右に行くに連れて、公開度や市民の参加度や関わり度は大きくなっている。

（Photo2）社会実験で行われた駅前地下広場における市民マーケット

さまざまな経緯を経て、2012年11月には「一般社団法人ひとネットワークひめじ」が設立され、2013年に緊急雇用により3名のスタッフが雇用された。2013年9月のサンクンガーデンおよび地下通路の先行オープンにより、本設の公共空間活用が「チャレンジ駅前おもてなし」として始まり、結婚式（Photo1）やスローハンモック、あおぞら大学、音楽ライブなど多彩な活動が展開されていった。活動数は徐々に認知度を高めるに比例して増加した。2014年7月の芝生広場のオープンとともに、活用可能な公共空間領域がさらに増えたことで、活動数は増加し、当時週末はほぼ活用枠が埋まり、月としても稼働率は50％を超えていた。初動期に主催団体とのネットワークの増加による活動や、杉活プログラム（姫路市産杉を活用した都市と農村をつなぐプロジェクト）の展開、マルシェ「えきまえ縁結び市」、「播磨べっちょない市」など定期開催の日常活動もプログラム化され、ネットワークや活動数増加の仕掛けをしてきた点は、成功要因の1つだと考えられる（Photo2）。これ以外に、日常的に高校生や子連れの主婦、ファミリー、高齢者が広場に集まり、日常的に、まちなかの居場所として、広場が息づいている。

（Photo1）社会実験で行われたサンクンガーデンにおける結婚式

> CHAPTER 2, 10 { CITIZEN PARTICIPATION AND AREA MANAGEMENT
Fostering Process of of Area Management
Special Contribution Rui Izumiyama

Area Management Development Process

The development of Himeji Station Square and the surrounding area in Himeji City, Hyogo Prefecture, proceeded through workshops and other citizen participation processes. In 2014, area management in Himeji City was still undergoing a process of trial and error with the efforts of the local government and related entities. The plaza in front of the station was not a complete area management project by itself; it was only a nucleus, a gateway to the city. Area management considering both the station and the city was just beginning. This section describes the process of physical development in front of Himeji Station Square, how the process was opened to the public, and how discussions and mechanisms for area management and the utilization of public space were established.

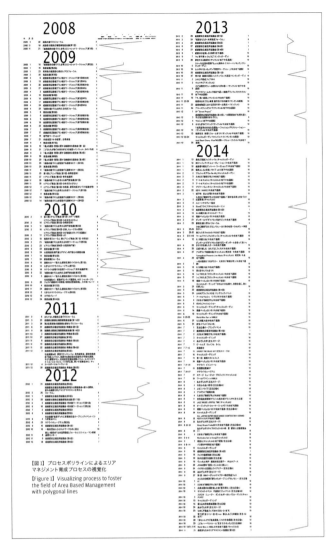

【図1】プロセスポリラインによるエリア
マネジメント育成プロセスの視覚化
【Figure 1】Visualizing process to foster
the field of Area Based Management
with polygonal lines

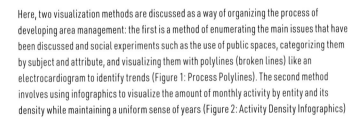

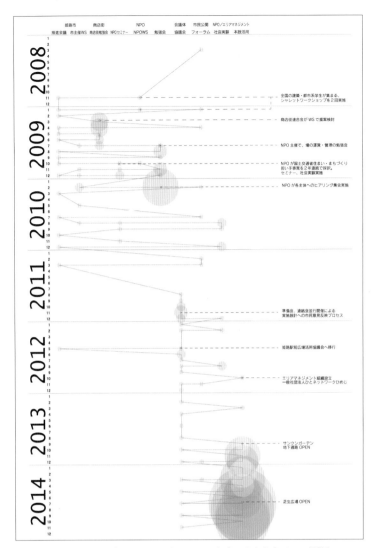

（Fig.1）プロセスポリラインによるエリアマネジメント育成プロセスの視覚化

（Fig.2）活動密度インフォグラフィックスによるエリアマネジメント育成プロセスの視覚化

Here, two visualization methods are discussed as a way of organizing the process of developing area management: the first is a method of enumerating the main issues that have been discussed and social experiments such as the use of public spaces, categorizing them by subject and attribute, and visualizing them with polylines (broken lines) like an electrocardiogram to identify trends (Figure 1: Process Polylines). The second method involves using infographics to visualize the amount of monthly activity by entity and its density while maintaining a uniform sense of years (Figure 2: Activity Density Infographics).

In both figures, from left to right: 1) Himeji City's official promotion meetings, 2) workshops organized by Himeji City, 3) study groups and workshops organized by the Federation of Shopping Centers, 4) seminars organized by the local NPO Slow Society Association, 5) workshops, 6) study groups, 7) conference bodies such as liaison groups and councils, and 8) public citizens. Looking from the left of the diagram, the degree of openness and citizen participation and involvement increases as one moves to the right of the diagram, from the city's official activities, to shopping districts, NPOs, conventions in which multiple entities gather, forums open to the public, and social experiments. As one moves to the right of the diagram, the degree of openness, participation, and citizen involvement increases.

After various processes, "Hitonetwork Himeji" was established in November 2012, and three staff members were hired in 2013 through emergency employment. The project began as "Challenge Ekimae Omotenashi," and a variety of activities such as weddings, slow hammocks, Aozora University, and live music performances were developed. The number of activities gradually increased in proportion to the growing awareness of the project, and with the opening of the lawn square in July 2014, the number of activities further increased as more public space areas became available for use. At the time, the weekend slots were almost completely filled and the monthly utilization rate exceeded 50%. In the initial stage of the project, activities through increased networking with the host organization, the development of the Sugi-Katsudo program (a project to connect urban and rural areas using cedar produced in Himeji City), and regularly held daily activities such as the Marche "Ekimae En-musubi Market" and "Harima Betsyonai Market" were also programmed to increase networking and the number of activities. This is considered to be one of the success factors. In addition, high school students, housewives with children, families, and the elderly gather at the plaza on a daily basis, making the plaza a place where people can live in the city.

Photo 1　A wedding ceremony in the Sunken Garden as part of the social experiment

Photo2　Citizens' market in the underground plaza in front of the station in as part of the social experiment

Fig.1　Visualization of area management development process by process polyline

Fig,2　Visualization of area management development process by activity density infographics

URBAN VISUALIZATION

CHANGES OUR TOWN

APPLICATION

CHAPTER 3

第三章　シミュレーションを生かしたまちづくりの実践

Chapter 3 1 登戸土地区画整理のためのデザインガイドライン
模型・景観シミュレーション・GISで将来ビジョンを見える化する

概要

区画整理事業によって大きくまちが変容する時に、地元住民に対し、地域の整備前・整備後の模型、交換テストによる景観整備のイメージ、GISによる地域情報などを眼に見える形で提示し、具体的な街並みのデザインガイドラインを策定した。また、地域で活動するNPO団体の拠点を利用者と共にデザインし建設した。

区画整理事業における合意形成

1988年の開始から2024年の現在まで36年という長期間を経過している神奈川県川崎市の登戸土地区画整理事業は、川崎市施工による公共事業で、道路などの基盤整備は2025年度に完了する。事業面積は約37.2ヘクタールと広大で、単なる農地の宅地化とは違って、すでに市街地になっている地区の施工であるため、建物の解体移転に膨大な時間がかかってしまった。私たちが関与した時期は、90年代の前半であるが、多くの建て替えが進むなかで、地域としての統一感が

求められた。そこで、市民ワークショップなどを繰り返し、地上7mで立面を分節するデザインガイドラインを策定した。

平成14年度	・基礎調査 ・権利者に対するアンケート調査の実施
平成15年度	・アンケート調査の分析 ・地理情報システム（GIS）による事業区域内の動向把握 ・協調的な街並み景観のためのルールづくりの検討 ・シミュレーションエリアの提案 ・ワークショップの実施
平成16年度	・住民参加によるワークショップの実施 ・住民の方々へのアンケート調査の実施 ・まちづくりガイドラインの提案

（Fig.1）景観ガイドライン策定のプロセス
アンケート → シミュレーション → 合意形成の流れが分かる

（Photo1）模型による確認風景

（Fig.2）CADによる交換テスト

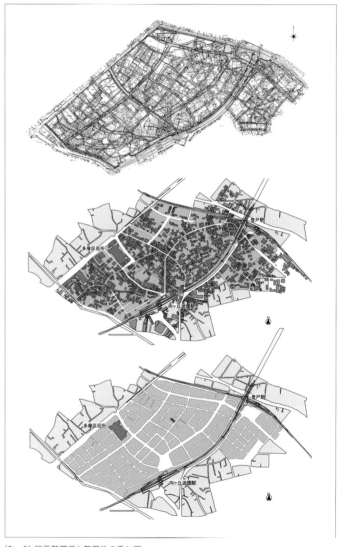

（Fig.3）区画整理前と整理後の重ね図

登戸駅前ガイドラインの概要

1. 低層部の連続性
   建物の低居部を7m程度の高さに揃えた、できるだけガラスを多用した開放性のある街並みにする。
2. 高層部のセットバック
   3階より上層の壁を低層部より最低50cmセットバックさせ、圧迫感を低減する。

3. カラーコントロール

建物の低層部の色彩をアースカラー（ベージュまたは茶系色）でまとめ、上層部と切り替える。上層部の色彩については激しい原色を避け、明度が高く、彩度を落とした色彩を使用する。

4. 植栽による緑の演出

歩道には出来るだけ植栽の場を確保し、潤いのある登戸地区のシンボル的な景観を演出する。

5. 看板の規制

巨大な屋上広告ではなく、適度な大きさの袖看板に統一する。

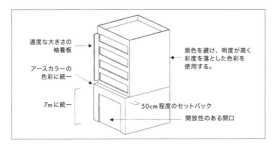

（Fig.4）登戸駅前ガイドラインのイメージ図

（Photo2）登戸駅前の現状
概ねヒューマンスケールが守られている

区役所通り商店街ガイドラインの概要

1. 低層部の連続性

建物の低層部を11m程度の高さに揃え、高層部をできるだけセットバックさせることで、圧迫感を低減させる。

2. 建物の壁面後退

看板、日除けテント、販売用のスペースを連続的に確保するため、建物の壁面線を道路境界線から 最低50cm後退させる。

3. 壁面線の連続性の維持

道路斜線緩和のために、建物の配置を道路からセットバックさせる場合は、道路沿いの壁面線の連続性を維持するような設えを考慮する。（ゲート、サイン、植栽など）

4. カラーコントロール

建物の低層部の色彩をアースカラー（ベージュまたは茶系色）でまとめ、上層部と切り替える。上層部 の色彩については、激しい原色を避け、明度が高く、彩度を落とした色彩を使用する。

5. 植栽による緑の演出

道路沿いやセットバックによる空地にはできるだけ植栽の場を確保し、潤いのある商店街の景観を演出する。

6. 看板の規制

巨大な広告禁止し、適度な大きさの袖看板に統一する。

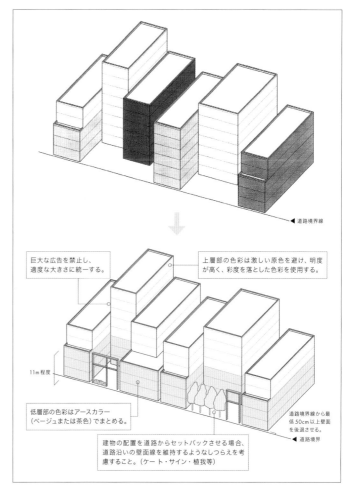

（Fig.5）区役所通り商店街ガイドラインのイメージ図

（Photo3）区役所通り商店街の現状
ガイドラインの強制力が弱いことはひとつの課題である

「NPO法人ぐらすかわさき」の
コミュニティフェ「遊友広場」の設計と施工

「NPO 法人ぐらすかわさき」から依頼され、新しい活動拠点「遊友広場」のあり方、デザインなどを学生のコンペなどを交えて議論し、合意形成をしながらデザインを進めた。最終的に、「まちへ開く広場」

（Photo4）デザインを決めるワークショップ

というコンセプトでまとまり、学生たちも工事に参加して、折れ戸によるファサードのあるコミュニティカフェを建設することができた。（現在は、区画整理事業のために移転している）

 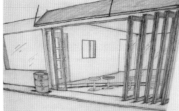

（Photo5）模型による空間イメージの提示　（Fig.6）スケッチによるデザインの改善

（Photo 6）完成時の様子

CHAPTER 3, 1 〉 DESIGN GUIDELINES FOR THE NOBORITO LAND READJUSTMENT PROJECT

"Visualization" of the Future of the Town Through Model, Landscape Simulation, and GIS

SUMMARY

When a town underwent a major transformation as a result of a land readjustment project, we presented the residents with models of the area before and after development, images of the townscape development through exchange tests, and local information through GIS in a visible form and formulated concrete design guidelines for the townscape. In addition, a base for a non-profit organization active in the community was designed and realized together with users.

The Noborito Land Readjustment Project in Kawasaki City, Kanagawa Prefecture, which has been in the works for 36 years since its inception in 1988, will be completed in FY2025, and the roads and other infrastructure will be built by Kawasaki City. The project area is vast (37.2 hectares), and unlike the mere conversion of farmland into residential land, the construction of this project took an enormous amount of time to dismantle and relocate the buildings because the area was already an urban area. We were involved in the project in the early 1990s, and while many rebuilding projects were underway, a unified streetscape was required for the area. Therefore, through a series of citizen workshops, we formulated a design guideline that divides the elevation at 7m above ground level.

1. Continuity of Low-rise Area

The low-rise portions of the buildings should be aligned at a height of about 7m, and as much glass as possible should be used to create an open and airy streetscape.

2. Setback of Upper Floors

Walls above the third floor should be set back at least 50cm from the lower floors to reduce the sense of oppression.

3. Color Control

The color scheme of the lower levels of the building should be earth colors (beige or brownish colors) and switched with that of the upper levels. Avoid intense primary colors in the upper levels of the building and use colors with high brightness and low saturation.

4. Greening by Planting

Planting areas should be secured along the sidewalks as much as possible to create a symbolic landscape of the Noborito district with a sense of richness.

5. Restrictions on Billboards

Unify signage with moderately-sized sleeve signs rather than huge rooftop advertisements (Fig.3).

Guideline for Kuyakusho-dori Shopping Street

1. Continuity of the low-rise area

The low-rise portions of the building are aligned at a height of about 11m, and the high-rise portions are set back as much as possible to reduce the sense of oppression.

2. Building Wall Setbacks

The building wall line should be set back at least 50cm from the street boundary line to provide continuous space for signage, sunshade tents, and sales areas.

3. Maintaining Continuity of Wall Lines

If the building is set back from the street to reduce the diagonal line of the street, consideration should be given to maintaining the continuity of the wall line along the street (gates, signs, plantings, etc.).

4. Color Control

Use earth colors (beige or brownish) for the lower levels of the building, and switch to earth colors for the upper levels. Avoid intense primary colors and use lighter, less saturated colors for the upper levels of the building.

5. Greening with Plantings

Planting space should be secured along the road and in setback areas to create a pleasant shopping district landscape.

6. Billboard Restrictions

Huge advertisements should be prohibited, and only moderately sized signboards should be used.

Design and Construction of "Yuyu Hiroba," a Community Café for "NPO Grass Kawasaki

At the request of the NPO "Grass Kawasaki," we discussed the design of the new activity center, "Yuyu Hiroba," for student competitions and proceeded with the design while building a consensus. In the end, the concept of a "plaza that opens to the town" was agreed upon, and a community café with a folding door façade was realized while students participated in the construction (the café has now been relocated due to a land readjustment project).

Fig.1 Process of formulating the landscape guidelines	Photo 3 Current status of Kuyakusho-dori shopping street
Fig.2 Overlaid view before and after rezoning	Photo 4 Workshop to decide on the design
Photo 1 Confirmation scene by model	Photo 5 Presentation of the space image using a model
Fig.2 Exchange test by CAD	Fig.5 Improving the design through sketches
Fig.3 Diagram of the guidelines in front of Noborito Station	Photo 6 Upon completion
Photo 2 Current situation in front of Noborito Station	
Fig.4 Image of the guideline for the Kuyakusho-dori shopping street	

横須賀市における眺望景観ガイドライン
眺望や景観を3Dモデルで見える化する

概要

わが国では、眺望景観を維持するための建築の高さ制限のルールはあまり多く展開されていない。私権にも関わる規制であるため、行政施策とするためには、十分に検討された根拠が必要となる。ここでは、久里浜湾からの眺望線を具体的に探り、眺望景観保全の基準とすることを試みた。主たる目的は、「合意形成を容易に進めていくためのツール開発」、「眺望景観に影響を与える開発に対する問題意識の共有」であったが、最終的には、本研究をもとにした景観基準案の妥当性が公的に認められ、横須賀市の「くりはま花の国眺望点」として制度化されている。

景観や眺望保全づくりの問題点

都市の高層化に伴い、今日の地方都市においては、特色ある山並みや海、ランドマークなどの自然景観への眺望が失われ、景観資源としての価値が下がるなど、眺望景観の悪化が問題となっている。新築される大規模建築物の高さや大きさを都市工学的、環境工学的な判断からの制限をかけていくだけではなく、その地域固有のコンテクストや景観資源を尊重し、それらへのインパクトを可能な限り軽減していくまちづくりのルールが、景観資源を有する地方都市において必要とされている。

眺望景観保全の具体的な取り組みとしては、1972年の横浜市における、「港の見える丘公園」からの眺望保全のためにふもとの地区の高さ規制を定めた要項がある。また、同じ頃に長野県松本市でも、国宝松本城周辺の隣接地でのマンション建設計画を契機として、眺望点・仰角が具体的に示された施策が打ち出され、これをもとに城周辺の建物高さの指導が開始された。さらに1984年には岩手県盛岡市での岩手山への眺望保全、1994年には岡山県の歴史的な後楽園・吹屋・閑谷3地区の眺望景観を保全するための周辺建物高さの規制実施などが挙げられる。

このように眺望保全に向けた自治体の取り組みは日本各地で増えてきている。しかし、景観資源に配慮されたルールをもつ自治体でも、景観や眺望保全づくりの基準方針として提供される情報では、一般の人からの理解が中々得られず、漠然とした方針のみにとどまっている。あるいは、眺望保全の必要性がうたわれていながら具体的な施策を取るに至っていないなど、合意形成上の問題が生じているのが現状である。その要因には、市民や事業者の景観への理解や認識を深めるための情報技術や市民レベルで検討可能なツールが自治体に乏しいこと、縦割り行政内の調整がうまく機能していない、などが挙げられる。

景観シミュレーションによる市民との合意形成

眺望景観保全への取り組みを進めるうえで、眺望景観を市民自身が考える機運をどうつくりあげ、地域の問題を市民レベルで検討できるようにすることが重要である。そのためには、①地域の街並みやコンテクストの把握（既存条件の整理）②開発に伴う問題意識の共有（インパクトの顕在化）③行政と市民の情報交換のためのシステム（プラットフォームの形成）の3点が必要である。3Dモデルを使った景観シミュレーションは、これらの条件を満たしていくうえで、有効なツールとしての期待が大きい。

具体的には、横須賀市行政と連携をはかり、「くりはま花の国眺望計画」を事例に横須賀市の地区計画のもとに、3Dシミュレーションを使った基準方針を作成し、行政内の意思統一や、市民合意を図るための「眺望景観保全のためのシミュレーション支援ツール」を開発した（Fig.1）。

（Photo1）「くりはま花の国」から眺望される久里浜湾の全景

（Photo2）ペリー提督が黒船から見たと思われる久里浜湾の風景

　まず、第一に、眺望景観保全基準づくりとして、空間的変化を与える高層建築が、眺望景観にどれだけの影響を及ぼし、全体のまとまりや街に相応しいかを、実際の目で確かめるための3Dシミュレーションツールを開発し、基準方針を決定した（Fig.2,3,4,5）。第二に、行政内部での基準報告会でシミュレーションを実装することによって、行政内部の意思統一や、眺望景観に対する意識がどのように変化したかを分析し、シミュレーションの有効性について検証した。第三に、市民への報告会において、眺望景観への意識がどうのように変化したかを分析し、シミュレーションが合意形成にどれだけの効果があるかという有効性を探ることが予定された（最終的に不開催となった）。

　これらのプロセスを経て、現在は、横須賀市のくりはま花の国眺望景観保全区域として制度化されている（Fig.6,7,8,9）。

シミュレーション支援ツール

ここではシミュレーション手順を示し、次章シミュレーションを行うこととする

（Fig.1）3Dシミュレーションを基にした景観基準方針の策定と実装するプロセス

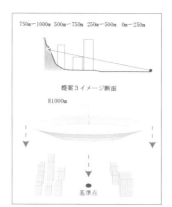

提案3イメージ断面

R1000m

基準点

（Fig.3）眺望景観検討図
久里浜湾の中心部の水面を視点場の基準点とし、その点を中心に、すり鉢状の眺望線を想定し、建物の高さ制限への基準につなげる

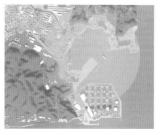

（Fig.4, 5）3Dシミュレーションによる眺望のイメージ
基準点から1kmの範囲を規制範囲とし、すり鉢状の眺望線を設定した

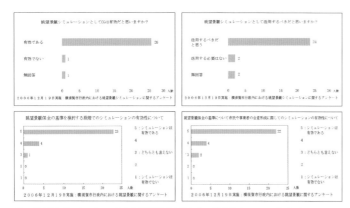

（Fig.2）3Dシミュレーションによる景観協議を経た後の行政体内内のフィードバック

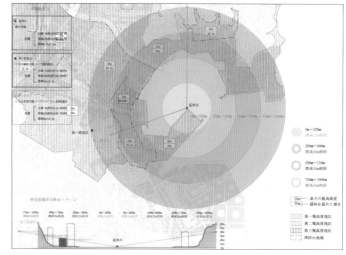

（Fig.6）くりはま花の国眺望景観保全基準

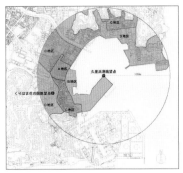 

(Fig.7)具体的な眺望景観規制図
（横須賀市HPより）

(Fig.8)具体的に高さ規制を示した表

眺望景観保全区域の解説（横須賀市）

「くりはま花の国」は、平成2年に策定された横須賀市都市景観整備基本計画で、眺望景観形成エリアとして位置付けされており、久里浜港内湾や東京湾、周辺の緑豊かな丘陵を望む良好な眺望の視点場が存在します。
また、久里浜港、内湾側からは、ペリーが見た久里浜の姿である居辺の緑豊かな丘陵への眺望があり、この眺望をペリー上陸の地として残しておくことも大切です。
これら久里浜ならではの眺望景観は、貴重な市民共有の財産といえることから眺望点を指定します。

(Fig.9)横須賀市による解説

Visualization of View Landscape with 3D Model

SUMMARY

In Japan, few rules for building height restrictions to maintain view landscapes have been developed. Since this regulation involves private rights, a well-considered rationale is necessary to make it an administrative measure. In this study, we attempted to specifically explore the view lines from Kurihama Bay and use them as a standard for preserving the view landscape. The main objectives were "to develop tools to facilitate consensus building" and "to share awareness of issues regarding development affecting view landscapes." Based on this study, Yokosuka City publicly recognized and institutionalized the appropriateness of the proposed landscape standards as "Kurihama Hananokuni Viewpoints."

Problems in Creating Landscape and View Preservation

As today's cities rise higher and higher, local cities are facing the problem of deteriorating scenic vistas, including the loss of views to distinctive mountain ranges, the sea, landmarks, and other natural landscapes, and a decline in their value as a scenic resource. It is necessary not only to restrict the height and size of new large-scale buildings from an engineering point of view but also to respect the unique local context and landscape resources and to create urban development rules that reduce the impact on these resources as much as possible.

One concrete example of an effort to preserve scenic vistas is Yokohama City's 1972 requirement to set height regulations for the foothill area to preserve the view from "Harbor View Park" in Yokohama. Around the same time, Matsumoto City in Nagano Prefecture also took the opportunity of a condominium construction project in the adjacent area around Matsumoto Castle, a national treasure, to come up with measures with specific viewpoints and elevation angles, and began to provide building height guidance around the castle based on these measures.

As described above, efforts by local governments to preserve views are increasing in many parts of Japan. However, even in municipalities with rules that consider scenic resources, the information provided as standard policies for creating scenic and view preservation is only a vague policy, as it is difficult for the general public to understand. In addition, although the need for view preservation has been mentioned, no concrete measures have been taken, resulting in consensus-building problems. The reasons for this include the lack of information technology and tools for citizens and businesses to deepen their understanding and awareness of the landscape and the lack of coordination within the vertically divided government.

Consensus Building through Landscape Simulation Using 3D Models

In promoting efforts to preserve scenic views, it is important to create momentum for citizens to think about scenic views themselves and to enable them to consider local issues at the citizen level. To achieve this, it is necessary to (1) understand the local townscape and context, (2) share awareness regarding the problems associated with development, and (3) create a system for information exchange between the government and citizens by forming a platform. The landscape simulation using 3D models is expected to be an effective tool in fulfilling these conditions. Therefore, in collaboration with the Yokosuka City government, we developed a "simulation support tool for view landscape preservation" to create a standard policy using 3D simulation.

First, a 3D simulation tool was developed to physically confirm the extent to which high-rise buildings that cause spatial changes affect views of the landscape and whether they are appropriate to the overall cohesiveness and townscape and to determine the criteria policy. Second, the effectiveness of the simulation was verified by analyzing how the implementation of the simulation in the internal debriefing session of the administration changed the internal consensus and their awareness of the viewscape. Third, we analyzed how awareness of the scenic view had changed during the debriefing session for the public and explored the effectiveness of the simulation in building consensus.

After these processes, the area is now institutionalized as the Kurihama Hananokuni View Landscape Conservation Area in Yokosuka City.

Explanation of the view landscape preservation area (Yokosuka City)
Kurihama Hananokuni is located as a scenic view formation area in the Yokosuka City Basic Plan for Urban Landscape Improvement formulated in 1990 and has a good viewpoint from which to view Kurihama Port, Tokyo Bay, and the surrounding lush green hills.
From Kurihama Port and the inner bay side, there is a view of Kurihama as Perry saw it, and it is important to preserve this view as the place where Perry landed.
The unique view of Kurihama is a valuable asset shared by the citizens of Kurihama and is designated as a viewpoint.

Chapter 3 — 3 岡山県高梁市における歴史的街並みの保存と再生
市民との協働プロセスを見える化する

概要

　約30年間にわたって、毎年岡山県の城下町に学生たちと数日間市内に滞在し、シャレットワークショップを実施して行政や市民から予め依頼されたまちの課題を解決するための方策の検討と提案を行った。歴史的街並みの調査、市民アンケート、まちの模型提示、景観の交換テスト、3Dシミュレーション、GISによる分析などを行い、市民との合意形成を図るとともに、解決策として誰にも分かりやすい具体的な提案を行った。その間、90年代の大学誘致と街並みの破壊、平成の大合併、「歴史まちづくり法」の認定、市立図書館と認定こども園の設計など、さまざまな事象が起きたが、結果としては、「都市の見える化」を通して、概ねまちの課題を一つひとつ解決の方向に導くことができた。

歴史的街並みの保存と問題点

　岡山県高梁市との出会いの契機は、知り合いの紹介から始まる。私は当時、ハーバード大学大学院での「ボストン市の保存と開発」研究から、帰国したばかりだったので、大学の研究室の活動として、まちづくり支援活動を快く受け入れる歴史的資源の豊富な地方自治体を探しており、1992年に初めて訪問した時に、ここしかないと運命的なものを感じたのである。

　高梁市の人口は90年代には約45,000人だったが、現在は約27,000人まで減少し、少子高齢化の波がますます強くなっている地方都市である。備中松山城の城下町として発展した高梁地区は、岡山県の西部を流れる高梁川の中流域と三方を山に囲まれた盆地に位置しており、2004年の市町村合併により、一市四町がまとまって一つの高梁市となった（Fig.1）。

（Fig.1）高梁市内の概略図
北の方から南に向けて、城下町が発展した

　地域の主な特徴と課題としては、次の3点があげられる。
（1）地形的に周囲が山に囲まれた盆地にあり、比較的境界が明確で平面的スプロールも少なく生活領域が把握しやすい。
（2）歴史的背景が豊かで、備中松山城のほかに武家屋敷町、寺町、町人町、高梁川を加え

（Photo1）備中松山城（国指定の重要文化財）

（Photo2）合併して高梁市に編入した吹屋地区
銅山によるベンガラの街並みが有名で、早い時期から伝統的建造物群保存地域に登録指定された

た都市修景要素が明確に保存されており、イメージアビリティが極めて高い。
（3）歴史的街並みの保存と世代交代や産業構造の変化による建替えの要求が相克し、放置しておくと確実に美しい街並みが破壊されることが予測された。

30年にわたる継続的な取り組みの概要

　私たちが初めて高梁市を訪問する直前の1990年から1992年まで、東京藝術大学の建築史家である前野まさる教授に指導され、伝統的建造物群保存対策調査（文化庁）が実施された。膨大な調査は各建物の寸法や意匠を克明に記録し、その後の「歴史的資産を生かした街づ

（Photo3）高梁市に関わる契機となった
学生用マンションの建設

くり」活動の基礎を形成した学術的価値の高いものだが、住民にその意図が十分に知らされなかったために、調査直後に伝統的町家のいくつかが壊され、学生用のマンションに建替えられるという不幸な出来事が起きてしまった。自由な建替えが制限される保存制度に対する市民の受け止め方は必ずしも好意的ではなかったのである。

　そのような状況下に飛び込んだ私たちの研究室活動の大きな目標は、市民の方々と話し合い、残存する街並みが得難い貴重な資産であり、保存制度を使わないまでも、出来できるだけ街の活性化に生かすことが地域固有の価値をの補強することに繋がるということを理解してもらうことであった。毎年、その強い意志を持って高梁市に通うことが始まった。

私たちの研究室の30年間にわたる、研究調査および提案の経緯については、内容的に大まかに6期に分けることができる。

1. 1993〜95年：基本的な情報収集期
2. 1996〜98年：住民との意見交換と方向模索期
3. 1999〜2001年：企画の実施とフィードバック期
4. 2002〜05年：状況変更による情報再収集期
5. 2006〜11年：第一次再編期
6. 2012〜23年：第二次再編期

それぞれの時期に実施されたシミュレーションとまちづくりの関係を見ていこう。

1. 1993〜95年：基本的な情報収集期

行政を窓口とした「おしかけ的アンケート調査」を市民、旅行者に対して行い、街並み景観を改良する提案を行った。広域を対象とした一般解を提案したので、ほぼ一方的な提案に終わる。

2. 1996〜98年：住民との意見交換と方向模索期

本町地区の街並みについて厳密な調査・分析を行い、街並み整備のための助成金制度立ち上げの支援をした。青年会議所と商工会議所を窓口に若い市民達との活動ネットワークを確立し、共同で「無電柱化」、「石畳による景観整備」、「門」の企画設計、「蔵」再生のビジョン、「歴史に残る人物記念館」の企画などを打ち出す。

（1）石畳と無電柱化の景観シミュレーション

街並み景観の交換テストをさまざまな街並みで実施した。（Fig.2）は、景観の交換テストにより、紺屋川沿いの舗装の変化と無電柱化の効果を示した時の図版である。このようなビジュアルを複数用意して住民の合意形成を図り、行政の予算化への流れをつくる支援をした。

（Fig.2）景観の交換テスト（左が整備前、右が整備後）
アスファルト舗装を石畳みに替え、無電柱化したイメージを比較している

（2）「門」のプロジェクト

歯抜けのようになっている駐車場の軒を連続的に修復するため、学生のデザインコンペによって選ばれたデザイン案を3D

（Photo4）「門」の整備前と整備後のイメージ（左が整備前、右が整備後）

化し、「門」のプロジェクトとして、青年商工会議所の費用で建設した。本町における景観修復の一つのモデルになっている（Photo4）（Fig.3）。

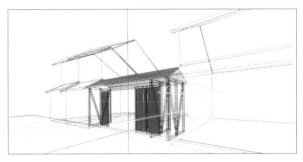

（Fig.3）事前に合意形成を図るための「門」の3Dイメージ

3. 1999〜2001年：企画の実施とフィードバック期

一度民間に渡って解体の危機を迎え、その後設立された「まちづくり会社」によって買い戻された「蔵」の改修工事を監修し、「まちづくり」の基地となる「高梁観光物産館」として再生することを支援した。また、市に強く要望していた「歴史的街並み保存地区整備計画」が策定され、本町地区を重点保存地区とし、街並み修復工事については建て主に対し助成金が支援されるようになった。そのために「街並みの遺伝子」をテーマにディスカッションを行い、具体的なデザインコードを住民と協議策定した。その後、伊達邸、小池邸、太田邸の修復工事の支援をした。

（1）「蔵」の再生プロジェクト

（Photo5）は、解体されそうになった蔵のイメージと、再生され地域の物産館となった「蔵」のプロジェクトのイメージである。蔵の実測、修復後の3Dイメージ作成、模型作成などにより、丁寧に市民間の合意形成を図った。

（Photo5）「蔵」の整備前と整備後のイメージ

（2）助成金制度と「高梁市のデザインガイドライン」の策定

デザインガイドラインとは、複数の建物群が無秩序に建設されることを予防するために、改修工事検討時に参考にすべき建物の外装材や部位のディテール、色彩などを決めた共通ルールのことをいう。高梁巾では、歴史的景観を維持するために、1998年に修景のための補助金を準備し、このデザインガイドラインに従って改修工事を行うときにのみ、補助金を支給することとした。このことにより、建物の持ち主の歴史的街並みへの意識を高めてもらい、時間をかけて連続性が途切れた街並みを修復することが意図された。まずは本町地区を重点保存地区として指定し、「歴史的街並み保存地区整備事業」を始め、その後対象地区は紺屋川地区、武家屋敷通りまで拡張されている（2024年現在）。

これに対応し、1997年までに私たちの研究室が行った本町地区の街並みに対する厳密な調査と分析に基づき、住民および行政との協議を行い、今後の街並みの修復工事で従うべきデザインガイドラインの策定を行った。(Fig.4)は街並みに関する調査結果の一部である。

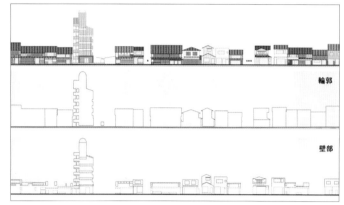

(Fig.4)街並みの構成要素の抽出(本町)
街並みを数段階のレアーに分解して要素間の関係を考える

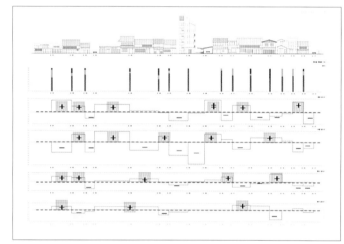

(Fig.5)街並みの外装材に関する分析グラフ

デザインガイドラインの具体的な内容については、歴史に裏付けされた伝統的な建築要素に対し、新しい遺伝子的要素を組み込む可能性について議論をした。その結果、旧来の街並みが示す色彩範囲に対し、より明るい生成りの木色を使用可能とし、現代という時代性を街並みに盛り込むことを試みた。

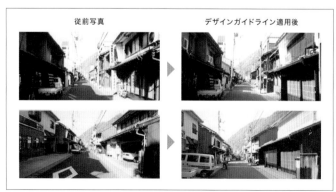

(Photo6)街並み修復の助成金制度による本町の街並みの変化

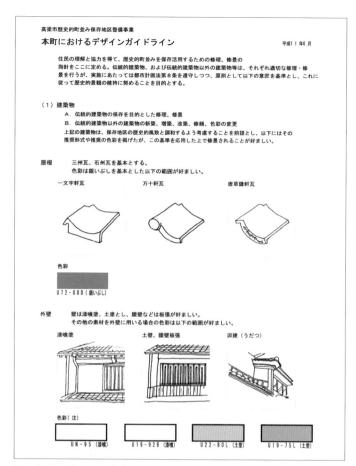

(Fig.6)屋根と壁に関するデザインガイドライン

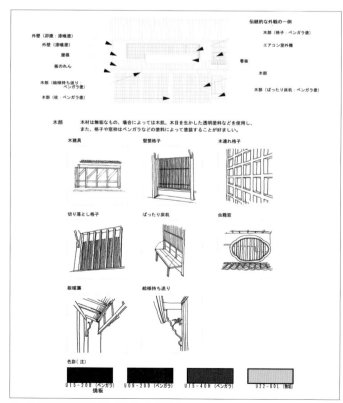

(Fig.7)格子戸などに関するデザインガイドライン

（Photo6）は本町地区の改修前と改修後の景観比較を行ったものであるが、明らかに街並みの連続性という点で 修景効果が認められたため、対象地区内の住民にも事業の意味が理解され、徐々に改修工事が着手された。このデザインガイドラインの策定以降約25年が経った2024年現在、すでに78軒がこのルールに沿って改修を実施し、歴史的街並みの連続性の回復に貢献している。

（Fig.8）はこの助成制度を利用して修復した伊達邸である。丁寧に、3Dや景観シミュレーションを行い、建て主、行政のコミュニケーションを促進することが重要である。

4. 2002〜2005年：状況変更による情報再収集期

2004年に近隣の4町（旧有漢町、旧成羽町、旧川上町、旧備中）を合併したため、行政単位が広域となり、今まで成羽町に所属していた吹屋町の重要伝統的建造物群保存地区が高梁市に移管されることになった。広域高梁市の街づくりも含め、今までの市内活性の取り組みだけでは対応できない問題が多く生まれたため、新たに調査を行い、街の情報を再収集した。

（Photo7）現役小学校として
100年以上使われた吹屋小学校
（Photo8）詳細な街並み調査の実施

5. 2006〜2011年：第一次再編期

高梁市の中心市街地を対象に包括的な景観データベースを作成し、街全体の情報についてGISを使ってデータ化した。また、特定の建物である旧守内邸および伊吹邸の再生計画を立てた。携帯電話を用いたユビキタス技術による情報発信の実験を始め、観光都市としての環境整備の方針を追求した。市長も変わって新たな地域再生の方向性が打ち出されたため、2010年には今までの調査研究の蓄積を生かし、国の「歴史まちづくり法」の認定を受けることができた。

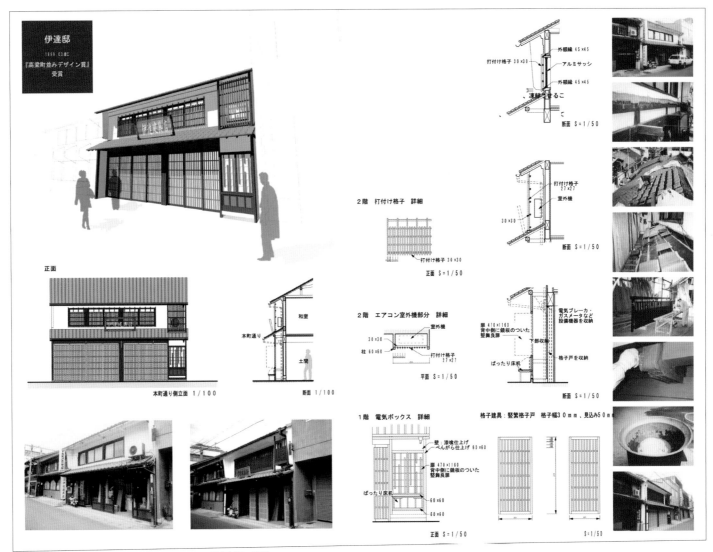

（Fig.8）伊達邸の外観に関するスタディーとそのための詳細断面の検討

（1）「歴史まちづくり法」の認定

　歴史まちづくり法（地域における歴史的風致の維持及び向上に関する法律）は、国土交通省、文化庁、農林産業省の三省にまたがった補助金制度で、人々の活動と建造物を包括する概念として「歴史的風致」を位置づけているところに特徴がある。祭礼行事などの人々の生活の営みへの支援に加え、その背景となるべき街並みの保全にも補助金が付くところが珍しい点である。高梁市と共に申請書をまとめ、全国の中でも早く認定を受けることができた（Fig.9）。

（Fig.9）歴史まちづくり法の申請資料

（2）旧守内邸の保存再生

　旧守内邸は本町通りの北部地区にあり、小路との角に位置している。2009年に、高梁市と明治大学が連携してこの建物を利用し、伝統的町家建築の構造・環境調査を実施した。

　その調査内容を踏まえたうえで、大学のサテライト研究室として利用する想定で保存再生の計画を行った。既存の建物を補強しながら、教育をテーマに高梁の街づくりを発信する場として有効的に活用することがこの計画の骨子であった。
　最近、横浜から高梁市に移住された横山夫婦が、ほぼ同じテーマでこの建物を使いたいというご意向を示され、この建物を購入された。その後、歴史的街並み保存地区のデザインガイドラインに則った外観の修復デザインを行い、補助金による保存再生工事が実施され、ほぼ完成した（2024年現在）。最初に調査した時から約14年が経過したが問題なく使用されており、伝統的町家建築の見本となる保存改修事例となっ

（Photo9）旧守内邸（整備前）

（Photo10）旧守内邸（整備後）

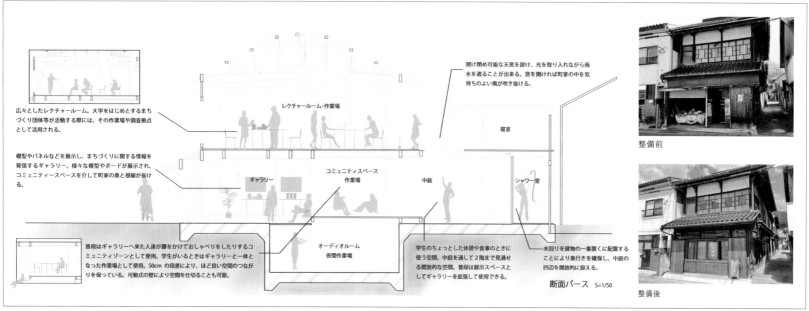

（Fig.10）当初の計画の断面図

た。(Photo9)と(Photo10)は修復前と保存再生後の外観の違いを示している。(Fig.10)は当初の計画案であるが、最終的には構造方式などが変更されたものの、大筋の考え方は継承された。現在は、中高生たちが学び交流する場として使われ始めている。

6. 2012年〜2023年：第二次再編期

(1) 立地適正化計画による居住誘導区域の設定

国土交通省がコンパクトシティーの政策を推進するため、「都市再生特別措置法に基づく立地適正化計画」という法律を定め、高梁市は2022年に策定を行った。これは、公共交通の駅を中心とした区域に医療、福祉、商業施設、住宅などを集約し、高齢者や子育て世代にとって安心できる快適な生活環境を確保するという意図である。そのコンパクトシティー構想

に則って、備中高梁駅の東西駅前ロータリーの整備、駅横図書館の建設、新しい認定こども園の整備などの話が動きはじめたため、それらの基本構想策定をサポートとして模型や3Dイメージを示す市民ワークショップを開催し、各々の整備計画の具体的な支援を行った。また、2023年秋には、約30年にわたる「歴史を生かしたまちづくり」が海外から評価され、私たちの研究室と高梁市が共同で「アジア都市景観賞」を受賞した。

(2) 高梁市立図書館の整備

私たちの研究室が基本構想を策定し、県内の設計事務所が実施設計を行い、2017年に竣工した。人口27,000人の町に「蔦屋」と「スターバックス」というブランドが指定管理者として入居したことが大きな話題になっている。基本構想の考え方がそのままの形で実現しているのが分かる。

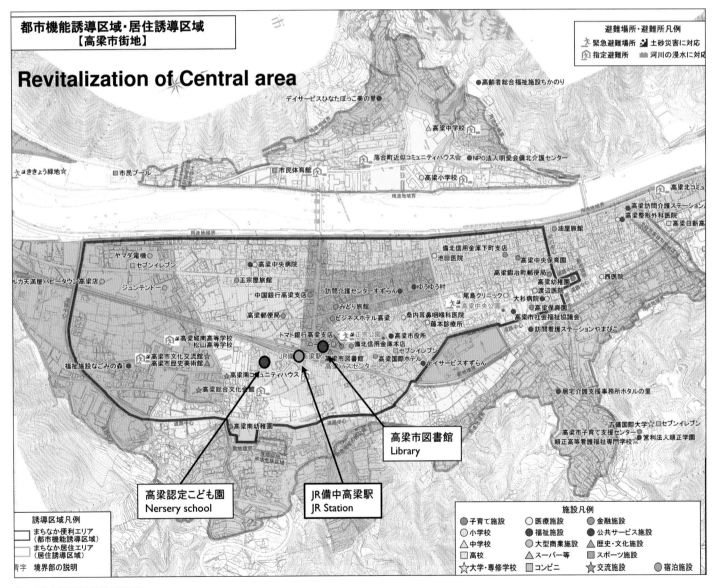

(Fig.11) 立地適正化計画の概略図
備中高梁駅の近くに、市立図書館が建設され、認定こども園の建設工事も開始された

（Fig.12）備中高梁駅西側ロータリーと図書館の鳥瞰イメージ

（Fig.13）備中高梁駅入り口のイメージ

（Photo11）竣工した図書館の外観とロータリー

（Photo12）竣工した図書館の内観

（Photo13）シャレットワークショップでの基本構想の提案

（3）梁認定こども園の整備

現存の幼稚園などを整理して、大規模な子ども園を駅の近くに建設する構想が市から提示されたため、2020年に研究室のシャレットワークショップの中で基本構想案を提示した。その後、毎年計画案を収斂しながら、市民ワークショップを繰り返し、基本計画・基本設計・実施設計を進めた。外観については、高梁市の景観計画に従い、歴史的街並みを意識した色彩を使用している。現在（2024年）は工事中で、2025年度に竣工予定である。

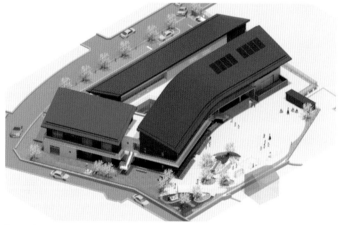

（Fig.14）高梁認定こども園の鳥瞰イメージ

（Fig.15）高梁認定こども園の鳥瞰イメージ

詳細は、『歴史的街並み再生のデザイン手法』（2013年、エクスナレッジ）参照

APPLICATION

Visualization of the Process of Collaboration with Citizens

SUMMARY

For about 30 years, every year, we stayed with students in a castle town in Okayama Prefecture for several days and conducted charrette workshops to study and propose measures to solve town issues as requested by the administration and citizens. We surveyed the historical streetscape, conducted a questionnaire survey with citizens, presented a model of the town, conducted a townscape exchange test, 3D simulation, and GIS analysis to build consensus with the citizens, and made concrete proposals for solutions that were easy for everyone to understand. During this period, various events occurred, such as the attraction of universities in the 1990s and the destruction of the cityscape, a series of large-scale municipal mergers carried out during the Heisei era, the approval of the "Historical Town Development Law," and the design of the municipal library and certified childcare center. However, through "urban visualization," we were generally able to lead the city's issues one by one toward solutions.

Preservation of Historical Streetscapes and Problems

Takahashi City is a regional city with a declining population of approximately 27,000, (compared to approximately 45,000 in the 1990s), and is experiencing an ever-growing trend of declining birthrates and aging population. The Takahashi area, which was developed to become the castle town of the Bitchu Matsuyama Castle, is located in a basin surrounded by mountains on three sides and the middle reaches of the Takahashi River flowing through the western part of Okayama Prefecture.

The region has the following three main characteristics.

(1) Topographically, the area is located in a basin surrounded by mountains, with relatively clear boundaries, making it easy to grasp the area where people live.
(2) The area has a rich historical background, and in addition to Bicchu Matsuyama Castle, the urban landscape elements of the samurai residence town, temple town, merchant town, and Takahashi River have been preserved, making the imageability of the area extremely high.
(3) The preservation of the historical townscape conflicted with the demand for reconstruction due to changes in the industrial structure, and it was predicted that the beautiful townscape would be destroyed if left unattended.

Overview of 30 years of Continuous Efforts

From 1990 to 1992, before our first visit to Takahashi City, a survey of measures by the Agency for Cultural Affairs to preserve traditional buildings was conducted under the direction of Professor Masaru Maeno, an architectural historian at Tokyo University of Fine Arts. The massive survey recorded measurements and designs of each building and was of great academic value, forming the basis for subsequent activities to create towns that make the most of their historical assets. However, unfortunately, citizens did not necessarily have a favorable attitude toward the preservation system that restricted freedom of reconstruction.

The main goal of our laboratory's activities was to discuss with citizens and help them understand the value of historical townscapes as assets and that they can be used to revitalize the town as a unique character of the area. With this strong will, we began to visit Takahashi City every year. The 30-year history of research and proposals can be

roughly divided into six phases.

1. 1993-1995: Basic information gathering
2. 1996-1998: Exchanging opinions with residents and seeking direction
3. 1999-2001: Planning implementation and feedback
4. 2002-2005: Re-gathering information due to changing circumstances
5. 2006-2011: First restructuring
6. 2012-2023: Second restructuring

1. 1993-1995: Basic information gathering

A "push survey" was conducted with citizens and tourists, with the government as the point of contact, to propose improvements to the cityscape.

2. 1996-1998: Exchanging opinions with residents and seeking direction

We conducted a rigorous study and analysis of the streetscape of the Honmachi district and assisted in launching a subsidy program to improve the streetscape. The Junior Chamber of Commerce and the Chamber of Commerce and Industry were the contact points for establishing a network of activities with young citizens, and they jointly launched a plan to promote no power poles, develop the landscape with cobblestone pavement, plan and design a "gate," create a vision for revitalizing the "warehouse," and plan a "memorial hall" for people who remain in the history.

(1) Simulation of the landscape with cobblestone pavement and no power poles

A townscape exchange test was conducted in various streetscapes (Fig.2). Figs.2 and 3 illustrate the effect of the change in pavement along the Kouya River and the conversion to no power poles.

(2) "Gate" project

To restore the parking lot eaves that destroy the continuity of the eaves, a design proposal selected through a student design competition was converted to 3D and realized as the "Gate" project at the expense of the Junior Chamber of Commerce.

3. 1999-2001: Planning implementation and feedback period

We supervised the renovation of a "warehouse" that was once in danger of being demolished by the private sector and then bought back by a "town development company" that was subsequently established. We assisted in its restoration as the "Takahashi Tourism and Products Museum," which would serve as a base for town development. In addition, the "Historic Townscape Preservation District Development Plan," which the city had requested, was formulated, and the Honmachi district was designated as a priority preservation district, and subsidies were provided to the builders for the restoration work of the townscape. Discussions were held on the theme of "townscape DNA," and specific design codes were discussed and formulated with the residents. Subsequently, we provided support for the restoration work at the Date, Koike, and Ota residences.

(1) "Storehouse" Restoration Project

Photos 5 and 6 show an image of a storehouse that was about to be demolished, and the restored "storehouse" project that turned into a local-product center. The project involved carefully building consensus among citizens by measuring the warehouse, creating a 3D image of the restored warehouse, and creating a model of the warehouse.

(2) Subsidy system and establishment of Takahashi City Design Guidelines

Design guidelines are common rules that determine the exterior materials, details of parts, and colors of buildings to be referred to when considering renovation work to prevent the uncontrolled construction of multiple groups of buildings. To maintain the historical landscape, Takahashi City prepared a subsidy for landscape improvement in 1998 and decided to provide the subsidy only when renovation work was carried out in accordance with these guidelines. The intention was to raise building owners' awareness of the historic streetscape and to restore the broken continuity of the streetscape over time.

In response to this, based on a rigorous survey and analysis of the Honmachi district streetscape conducted by our laboratory in 1997 (Fig.4), we held discussions with the residents and the government and formulated a design code to be followed in future streetscape restoration work.

Photo 7 shows a comparison of the landscape of the Honmachi district before and after the renovation. The project had a positive effect on the continuity of the streetscape. As of 2024, approximately 25 years after the establishment of these design guidelines, 78 houses have been renovated in accordance with these rules, contributing to the restoration of continuity in the historic streetscape.

Fig.8 is the Date residence that was restored using this grant program. It is important to carefully conduct 3D and landscape simulations to promote communication between the builder and the government.

4. 2002-2005: Information re-collection due to change in situation

In 2004, four neighboring towns were merged, resulting in a wider administrative unit, and the Important Preservation District for Groups of Traditional Buildings in Fukiya Town, which had previously belonged to Nariwa Town, was transferred to Takahashi City. Many issues arose that could not be addressed by the existing efforts to revitalize the city alone, including town development in the wider Takahashi City; therefore, a new survey was conducted and information on the town was re-collected.

5. 2006-2011: First restructuring

A comprehensive landscape database was created for the central city area of Takahashi City, and information on the entire city was databased using GIS. In addition, a revitalization plan for specific buildings, the former Moriuchi Residence and the Ibuki Residence, was developed. The city began to experiment with information dissemination through ubiquitous technology using cell phones and pursued a policy of improving the environment as a tourist city. The mayor changed and a new direction for regional revitalization was set, and in 2010, the city was able to receive government approval for the "Historical Town Development Law" by taking advantage of the accumulation of studies conducted to date.

(1) Certification under the Historic Town Development Law
The Historic Town Development Act is a three-chapter subsidy system under the Ministry of Land, Infrastructure, Transport and Tourism, the Agency for Cultural Affairs and the Ministry of Agriculture, Forestry and Industry and is unique in that it positions "historic properties" as a concept that encompasses people's activities and buildings. It is unique in that it provides subsidies not only for festival events and other activities of people's daily lives but also for the preservation of the townscape that supports such activities.

(2) Preservation and Restoration of the Former Morinouchi Residence
In 2009, Takahashi City and Meiji University collaborated to use this building to conduct a structural and environmental survey on traditional machiya architecture. Based on the survey, a plan for conservation and rehabilitation was developed on the assumption that the building would be used as a satellite laboratory for the university. While reinforcing the existing building, the plan's framework was to effectively utilize the building as a place to disseminate Takahashi's urban development under the theme of education.

A Yokoyama couple, who recently moved from Yokohama to Takahashi, expressed their desire to use this building for almost the same theme and purchased the building. Subsequently, the exterior of the building was designed for restoration in accordance with the design guidelines of the Historic Townscape Preservation District, and subsidized conservation and rehabilitation work was carried out, which is about to be completed (as of 2024). The building has been in use without any problems for about 14 years since the time of the initial survey and has become an example of conservation and renovation that serves as a model for traditional townhouse architecture. Currently, the building is being used as a place for middle and high school students to learn and interact.

6. 2012-2023: Second restructuring

(1) Establishment of residential guidance zones based on the Location Optimization Plan
To promote the compact city policy, the Ministry of Land, Infrastructure, Transport and Tourism established a law called the "Location Optimization Plan based on the Act on Special Measures for Urban Revitalization" and Takahashi City formulated the plan in 2022. The intention was to secure a safe and comfortable living environment for the elderly and the child-rearing generation by concentrating medical care, welfare, commercial facilities, housing, etc. in areas around public transportation stations. In accordance with the compact city concept, the development of the east-west station roundabout at Bicchu Takahashi Station, the construction of a library next to the station, and discussions about the development of a new certified children's kindergarten began. In addition, in the fall of 2023, the company held a workshop for the public to show models and 3D images. In the fall of 2023, our laboratory and Takahashi City jointly received the "Asian Urban Landscape Award" in recognition of their "city planning utilizing history" over a period of about 30 years from overseas..

(1) Development of Takahashi Municipal Library
Our laboratory developed the basic concept, a design firm in the prefecture carried out the implementation design, and construction was completed in 2017. The moving in of the brands "Tsutaya" and "Starbucks" as designated managers in a town with a population of 27,000 has become a major topic of discussion. It can be seen that the concept of the basic plan has been realized in its original form.

(2) Development of Takahashi Certified Children's Kindergarten
The city proposed the concept of constructing a large-scale children's garden near the station by organizing existing kindergartens and other facilities, and we presented a basic concept plan during the charrette workshop in 2020. Since then, the basic plan, basic design, and implementation design have been carried out through repeated citizen workshops while converging the plan drafts every year. For the exterior, colors were used in accordance with the Takahashi City landscape plan, with an awareness of the historical townscape. Construction is currently underway (2024) and is scheduled to be completed by 2025.

Fig.1　Schematic diagram of Takahashi City
Photo 1　Bicchu Matsuyama Castle (National Important Cultural Property)
Photo 2　Fukiiya District, which was incorporated into Takahashi City after the merger of the two cities
Photo 3　Construction of the student apartment building that led to our involvement with Takahashi City
Figs.2　Townscape exchange test before (left) and after (right)
Photo 4　Images of the "Gate" before (left) and after (right) restoration
Fig.3　3D image of the "Gate"
Photo 5　Images of the "storehouse" before and after restoration
Fig.4　Extraction of the components of the streetscape (Honmachi)
Fig.5　Analysis graph of the exterior materials of the streetscape
Fig.6　Design guidelines for roofs and walls
Fig.7　Design guidelines for lattice doors, etc.
Photo 6　Changes in Honmachi's streetscape as a result of the subsidy program for streetscape restoration
Fig.8　Study on the exterior of the Date Residence and detailed cross section
Photo 7　Suiya Elementary School, used as an active elementary school for more than 100 years
Photo 8　Conducting a detailed town survey
Fig.9　Application materials for the Historic Town Development Act
Photo 9　Former Mooriuchi Residence (before maintenance)
Photo 10　Former Moorouchi Residence (after maintenance)
Fig.10　Schematic diagram of the location optimization plan
Fig.11　Cross-section of the original plan
Fig.12　Bird's-eye view of the west side rotary of Bicchu Takahashi Station and the library
Fig.13　Image of the entrance to Bicchu Takahashi Station
Photo 11　Exterior of the completed library and the rotary
Photo 12　Interior view of the completed library
Photo 13　Proposal of the basic concept at the charrette workshop
Fig.14　Bird's-eye view image of Takahashi Certified Children's Garden
Fig.15　Interior view of the Takahashi Certified Preschool

*For details, see "Design Methods for Historical Townscape Revitalization" (2013, X-knowledge)

Chapter 3 / 4 東京都下北沢地区の道路問題とまちづくり支援
「対立」から「対話」への転換

概要

　東京 下北沢の都市計画問題を契機に、市民活動を支援する形で、さまざまな「見える化」ツールを駆使して、関係者間の合意形成を促進した事例。当初は、都市計画道路事業が施行された「まちの変容」予測を模型と3Dシミュレーションで示し、感情論による道路反対運動にしない方向を探った。その後、小田急線地下化上部利用のビジョンを市民と共に育てる経過において、シャレットワークショップを何回も開催し、GISによる地域の情報分析を駆使した。最終的には、さまざまな局面で示したビジュアルが、実際の開発ビジョンを先導し、行政・鉄道事業者・市民による理解が深まり、心地よいウォーカブルな都市空間の創造に貢献することができた。現在、その経過で育った「緑」を中心にした市民グループが活発に地域活動を続けている。

　世田谷区下北沢地区は、東京の都心西部には珍しく、ヒューマンスケールの路地のある街や演劇・音楽のメッカとして若者に愛される魅力的な界隈として有名である。一般には、「下北沢地区の再開発」として世間では知られているが、下北沢のまちづくりの変遷過程が、(1)連続立体交差事業(小田急線の複々線事業を含む)を起因とした都市計画道路問題 (2)それらの結果としての小田急線上部利用計画 (3)その後の醸成された市民参加によるまちづくりなど、いくつかの要素が複合したものとなっていることはあまり知られていない。

　私たちの研究室は、都市計画道路計画が初めて住民や市民に示された2003年ころから、この計画の問題に関わることになった。さまざまな経緯があり2024年時点で約20年の年月が経過したが、ほぼ「街の変化」の全体像が見えてきたところである。最近では、小田急線が地下化された上部に緑豊かな空間が生まれ、テレビドラマのロケが行われるくらいに注目される都市空間が誕生した。

　これまでのタイムフレームをあまり知らない世代も多いと思われるので、ここでは今までの経緯を振り返った後、下北沢のまちづくりの問題が私たちの社会に与えた意味についても触れてみる。

　下北沢のまちづくりの問題は、大きく以下の三期に分けることができる。

一期：都市計画道路補助54号線をめぐる紛争期
(2003〜2006年)

　一期においては、連続立体交差事業に伴う広幅員の都市計画道路が下北沢地区の中心部を横断することについて、市民と行政の間で厳しい対立が生まれ、ニューヨーク・タイムズにまで取り上げられるほどの深刻な紛争が広がった。そもそもの発端は、交通渋滞を生む踏切をなくすための連続立体交差事業、(国、都、事業者による費用分担)の必要条件に二本の幹線道路との交差が規定されていたことから、1946年に都市計画が決定されたが事業化されずにいた都市計画道

(Photo1) 模型による将来像の予測

路補助54号線(幅員26m)を改めて事業化する方針を、世田谷区が2003年に打ち出したことから始まる。これと並行して、下北沢地区に新たな地区計画をかけ、下北沢特有の狭小道路においても中高層の建物が建てられる構想が区から提案された。(2000m²以上の敷地には、60mの高層ビルが建設可能)

　この時発表された計画のイメージがあまりに下北沢に合わないものであったので、多くの人々は衝撃を受け、意識の高い住民や下北沢を愛するミュージシャンたちが中心になって、大きな反対運動として展開した。一方、地域の専門家が介在し、単なる反対ではなく代替案による解決策を探ろうとする「下北沢フォーラム」が並行して組織され、市民との協働を推進する活動を展開した。私たちの研究室は、近隣の住民から依頼され、今回の都市計画道路、地区計画による下北沢の将来イメージをシミュレーションした。

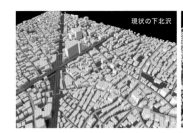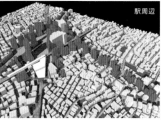

(Fig.1) 3D画像による都市計画道路によるインパクトの予測

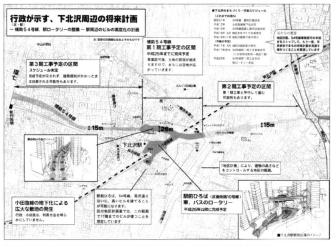

(Fig.2) 行政による情報を分かりやすく解説した画像の提供(荻原礼子氏作成)

その後、市民の意向を調査することが重要であると考え、「下北沢フォーラム」と研究室が協働で、3,000名に対する意向調査を行った。この結果、ほとんどの住民は、都市計画道路や高層ビルの街を望んでいないことが明らかになった。

（Photo2）アンケートの集計作業

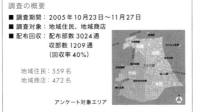

調査の概要
■ 調査期間：2005年10月23日〜11月27日
■ 調査対象：地域住民、地域商店
■ 配布回収：配布部数 3024通
　　　　　　収集部数 1209通
　　　　　　（回収率40%）

地域住民：559名
地域商店：472名

アンケート対象エリア

（Fig.3）アンケートの配布範囲

計画案についての認知と評価
大規模道路整備（補助54号線）について

不必要：住民で64%、商店で55%

必要：住民・商店ともに10%台

（Fig.4）アンケート結果の一部

代替案による具体的な将来像を模索するために、ハーバード大学、慶應義塾大学、明治大学の大学院の演習課題で、この問題を取り上げ、それぞれの大学院生が提案を行った。基本的には、道路は不要とし、駅周辺をウォーカブルな街にしようという意味では共通点があった。

（Photo3）ハーバード大学の学生による調査

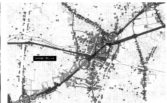
（Fig.5）ハーバード大学の学生による提案

（Fig.6）慶應義塾大学の学生による提案

（Fig.7）明治大学の学生による提案

広場のイメージ　B案
（Fig.9）専門家による代替案の提案（駅前広場の将来イメージ）

この調査結果を世田谷区に伝達し、計画見直しを要望したが、変更の方向には動かなかった。そのため、道路計画反対運動はさらに激しくなったが、「下北沢フォーラム」は専門家を集めた「シャレットワークショップ」を市民に公開する形で数回開催し、代替案を策定した。2006年の4月には、多くの学識者によるシンポジウムも開いて、世田谷区への提言を行った。

（Fig.8）専門家による代替案の提案（平面図）

京王井の頭線の法地（本多劇場の前）を一方通行の道路として整備し、ミニバスとタクシーなどの公共交通のみを駅前広場の周縁部に通す。

〈 凡例 〉
――　補助54号線
||||||||　主要道路

（Photo4）道路計画反対グループによるデモの風景

（Photo5）第1回目のシャレットワークショップ

((Photo6) 都市系専門家によるシンポジウム

（Photo7）世田谷区都市計画審議会における強行採決

二期：小田急線上部利用計画をめぐる黎明期
（2006〜2013年）

二期においては、都市計画道路問題のショックが冷めやらぬなか、数年後に小田急線の線路と世田谷代田駅、下北沢駅、東北沢駅が地下化され、その上部に広大なスペースが出現することが住民にも周知され、その空間利用が下北沢のまちづくりの新たなテーマとなった。新しく組織された「小田急線あとちの会」は世田谷区と小田急電鉄に対し、上部利用計画に市民のアイデアが反映されるような機会をつくることを進言し、2008年には世田谷区主催による「市民アイデアを募集するイベント」が開催されることにつながった。その後、東日本大震災の起こった2011年の春に世田谷区長選挙が行われ、震災時に基礎自治体の存在意義を体感し、情報公開と市民参加を全面的に打ち出した保坂展人氏が新たな世田谷区長に就任した。新区長は、防災的な観点から小田急線上部のスペースにはできるだけ建物を建てず、避難のための野外空間が必要であると提唱した。「小田急線あとちの会」が発展した「グリーンライン下北沢」という組織は、2012年に小田急線上部空間が緑豊かな野外公共スペースとして使われる具体的なイメージを作成して、世田谷区に対し提言をしたところ、新区長には街の将来ビジョンとして肯定的に受け入れられた。

（Photo9）学生によるシャレットワークショップ

（Photo10）保坂展人新世田谷区長の誕生

APPLICATION

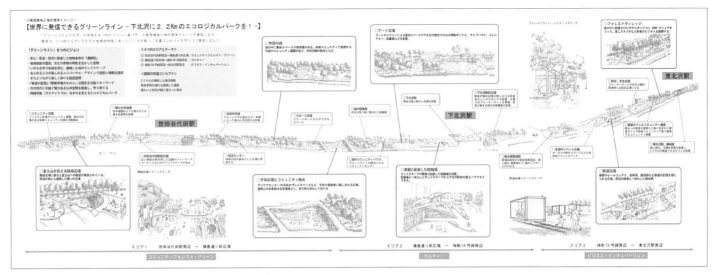

(Fig.10) NPO グリーンライン下北沢による小田急線地下化上部計画の提案
災害を予見し、基本的には建物は建てないグリーンコリドーを提案している

三期：まざまな市民参加のまちづくり活動と具体的な空間の出現期（2013年〜）

（Photo11）北沢デザイン会議の風景　（Photo12）北沢PR戦略会議の風景

　三期における最大の変化は、ステイクホルダー間の新たなコミュニケーションの実現である。当初は抵抗圧力があったが、世田谷区の街づくり課が「北沢デザイン会議」という市民、行政、事業者間のラウンドテーブル的な会合の枠組みを組み立て、2014年の8月から現在までに7回開催されている。これは、小田急線の上部空間のデザインおよび下北沢のまちづくりについての情報公開と市民参画の可能性を求めた催しであるが、都市計画道路問題の時期とは隔絶な違いとなっている。一方、変化する下北

（Fig.11）北沢デザイン会議と北沢PR戦略会議のちらし

沢の街や新しく整備される空間に積極的に関わる人材を発掘し、具体的な活動を推進する場として「北沢PR戦略会議」も民間主導で組織され、2016年から活動が始まった。この会議体には「緑部会」「駅広部会」「ユニバーサルデザインチーム」などが立ち上がり、市民が自分の関心あるグループに所属して、ソフト的活動に自発的に取り組むことができる仕組みが動き始めた。戦略会議は、現在までに報告会も含めて14回開催されており、それまで顔を出さなかった小田急電鉄も当初の対話拒否の態度から徐々に対話重視に変化し、市民の声に積極的に耳を傾けた「支援型開発」という開発方針を打ち出した。

　具体的な小田急線上部利用の空間デザインについては、大手設計事務所と私たちの研究室が共同して、建物の建てられ方について2014年に民間所用地を対象とした「デザインガイドライン」を作成した。ここでは、既存の街並みのスケールを建物のデザインに生かし、歩きたくなる空間をつくり出すビジョンを提案した。その結果、あまり重い建物が建てられない小田急線の軌道上の土地には、「BONUS TRACK」に代表される小スケールの建物が分節して建てられ、下北沢の売り物であるヒューマンスケールな街並みや広場が実現した。2.2kmにわたる小田急線上部空間には、公共用地と民間所有地の境界のない連続する緑空間が実現し、現在は市民による「シモキタ園藝部」というグループが管理を担当している。これに交えて、「大学のサテライトキャンパ

（Photo13）新しく誕生した緑道の風景

ス」「保育園」「温泉旅館」「学生寮」「店舗兼用住宅」「商業施設」「宿泊施設」などさまざまな用途の空間がリニアーに並び、国内でも例を見ない面白い散歩空間が生まれた。これらの施設は、訪問客と共に地域住民にも親しまれ、周辺居住環境におけるQOL（生活の質）を上げることに大きく貢献したことは間違いない。

（Photo14）BONUS TRACKの小スケールの建物と広場の風景

下北沢のまちづくりが現代に与える意味

　下北沢が今の日本の社会に与える意味については、大きく2つ挙げることができる。

　1つ目は、利用されなくなった都市の土木インフラを再利用し、新しい環境装置として都心型の「グリーンインフラ」を実現した事例としての価値である。もともと人が集まる広場や緑地が少なかった下北沢地区には生物多様性を考えるキャパシティーが乏しかったが、ニューヨークのセントラルパークを実現したF.L.オルムステッド（ランドスケープデザイナー）が計画した「エメラルドネックレス」という緑の連鎖が、現在ボストンの街の中で極めて重要な役割を果たしているように、小田急線上部の緑の連鎖が数十年先には、東京の重要な緑の帯として実を結ぶことが期待されている。現在注目されている「グリーンインフラ」は、広義的には防災性、資源循環、景観向上、コミュニティー維持などが含まれるが、下北沢のまちづくり事例では将来的に緑を媒介とした「グリーンコミュニティー」の熟成が期待されている。

　2つ目はさらに重要なものであるが、さまざまな経緯を経るなかで、市民たちが政治的対立構造を乗り越え、自ら参画協働するプロセスを学んで実践し、それが目に見える形で街の変化に反映された点である。当初、都市計画道路を巡った紛争時には、行政と市民の対立による分断など不安な状況が続いたが、区長が変わったことを契機に、ラウンドテーブル的な話合いの場やワークショップが継続的に開催され、行政、市民、事業者が同じプラットフォームで対話をはじめたことが大きい。一般の都市系まちづくりでは、平常時の政策に関する話し合いが淡々と行われることが想像されるが、下北沢では街の危機を経験することで、社会関係資本（ソシアルキャピタル）が加速的に醸成されたのではないかと思われる。「想い」の強い市民が積極的に緑や広場の運営に関わろうとしている姿は、意志あるステイクホルダーによる地域自治の可能性を期待させ、今後展開されるであろうエリアマネジメントの下地をしっかり形成されていることが分かる。

CHAPTER 3, 4 　ROAD PROBLEM IN SHIMOKITAZAWA DISTRICT OF TOKYO AND SUPPORT FOR TOWN PLANNING

From a Confrontation to a Dialogue

SUMMARY

This is an example of how an urban planning issue in one area, Shimokitazawa, Tokyo, prompted the use of various "visualization" tools to facilitate consensus among the parties involved in the form of support for citizen activities. Initially, a model and 3D simulation were used to show the projected "transformation of the town" in which the urban planning road project was to be implemented to find a direction that was not a campaign against the road based on emotions. Later, in the process of developing a vision for the use of the upper portion of the Odakyu Line underground with the citizens, we held several charrette workshops and made full use of GIS to analyze information about the area. In the end, the visual images presented at various stages of the project led to an actual development vision, which deepened understanding among the government, railroad operators, and citizens and contributed to the creation of a pleasant walkable urban space. Currently, a citizen group centered on the "greenery" that grew out of this process continues to be active in the community.

The Shimokitazawa district of Setagaya Ward is famous as an attractive neighborhood loved by young people as a town with human-scale alleys and a mecca for theater and music, which is rare in the western part of central Tokyo. Although generally known to the public as "the redevelopment of the Shimokitazawa area," the transition process of Shimokitazawa's town development has been influenced by (1) urban planning road issues resulting from the continuous multi-level crossing project (including the Odakyu Line double track project), (2) the Odakyu Line upper utilization plan resulting from these issues, and (3) the subsequent city planning with fostered citizen participation, etc. It is not well known that this is a combination of several factors.

Our laboratory has been involved in this planning issue since around 2003 when the urban planning road plan was first presented to residents and citizens. Due to various circumstances, as of 2024, about 20 years have passed, and we are beginning to get an overall picture of the changes in the city. Recently, a lush green space has been created above the Odakyu Line, which has been converted to underground, creating an urban space that has attracted so much attention that a TV drama has been filmed there. Since most of our generation may be unfamiliar with the timeframe up to this point, I will review the history of the project and then discuss the significance of the Shimokitazawa urban development issue to our society.

The Shimokitazawa urban development issue can be roughly divided into the following three phases.

Phase I: The dispute over the urban planning road Auxiliary Route 54 (2003-2006)

In this first period, a severe conflict arose between the citizens and the government over a wide urban planning road that crossed the center of the Shimokitazawa area in conjunction with a multi-level crossing project, and the dispute was so serious that it was even reported in the New York Times. The dispute began when Setagaya Ward decided to redevelop the 26m wide urban planning road Auxiliary Route 54, which had been approved in 1946 but not yet put into operation, because it was stipulated that the road must cross two arterial roads as a prerequisite for the continuous grade crossing project (cost sharing among the national government, metropolitan government, and the project operator) to eliminate railroad crossings that create traffic congestion. In parallel with this, the ward proposed a new district plan for the Shimokitazawa area, which would allow the construction of mid- to high-rise buildings (60m-high skyscraper can be built on a site of more than 2,000m2) even on the narrow roads characteristic of Shimokitazawa.

The image of the plan presented at the time was so uncharacteristic of Shimokitazawa that many people were shocked, and a large opposition movement led by conscious residents and musicians who loved Shimokitazawa emerged. At the same time, the "Shimokitazawa Forum" was organized with the intervention of local experts to seek not just opposition but alternative solutions and to promote collaboration with the public. Our laboratory was requested by neighborhood residents to simulate the future image of Shimokitazawa with this urban planning road and district plan. Subsequently, the "Shimokitazawa Forum" and the laboratory conducted a survey of 3,000 people, and the results revealed that most residents did not want urban planning roads and high-rise buildings in the city (Photo 2).

To seek a concrete vision of the future through alternative proposals, this issue was taken up in graduate school exercise assignments at Harvard University, Keio University, and Meiji University, where each student made a proposal. Basically, there were similarities in the sense that roads were unnecessary and that the station area should be converted into a walkable town.

The results of this study were communicated to Setagaya Ward and a request was made for a review of the plan, but no movement was made in the direction of change. The Shimokitazawa Forum held several "charrette workshops" open to the public, bringing together experts to develop alternatives. In April 2006, several academics held a symposium to make recommendations to Setagaya Ward.

However, the administration refused to acknowledge any of these requests, and finally, in October 2006, the Setagaya Ward Urban Planning Council approved the urban planning road Auxiliary Route 54 project and the district plan at the same time. The city planning decision was so undemocratic and uncharacteristic of Setagaya Ward that for a while after 2006, no one talked about the road issue. The group opposing the road plan failed to achieve its original goal and was effectively disbanded.

Phase II: The Dawn of the Odakyu Line Upper-Line Utilization Plan (2006-2013)

In the second phase, while the shock of the urban planning road issue was still fresh, residents became aware that the Odakyu Line tracks and Setagaya-Daita, Shimokitazawa, and Higashi-Kitazawa stations would be converted to underground several years later and that a vast space would appear above them. The newly organized "Odakyu Line Atochi no Kai" (Association of Upper use of the Odakyu Railway Track) advised Setagaya Ward and Odakyu Electric Railway to create opportunities for citizens' ideas to be reflected in the upper utilization plan, which led to the holding of an event in 2008 hosted by Setagaya Ward to solicit citizens' ideas. In the spring of 2011, when the Great East Japan Earthquake struck, an election was held to elect a new mayor for Setagaya Ward, Nobuto Hosaka, who had experienced the significance of the existence of a basic local government at the time of the earthquake and had been fully committed to information disclosure and citizen participation. The new mayor advocated that, from a disaster prevention perspective, as few buildings as possible should be constructed in the space above the Odakyu Line and that an open-air space for evacuation is necessary. In 2012, the organization Green Line Shimokitazawa, which grew out of the Odakyu Line Atochi no Kai, created a future image of the space above the Odakyu Line to be used as a lush outdoor public space and proposed it to Setagaya Ward, which was accepted positively by the new mayor as his vision for the city.

Phase III: Various citizen-participatory community development activities and the emergence of concrete spaces (2013-)

The biggest change in the third period is the realization of new communication among stakeholders. Despite initial resistance pressure, Setagaya Ward's urban development division has assembled a framework for roundtable-like meetings among citizens, government, and the Kitazawa Design Conference, which has been held seven times since August 2014 to date. Meanwhile, the "Kitazawa PR Strategy Council" was also organized

under the leadership of the private sector, which began in 2016 to promote specific activities in the newly developed space, and the "Green Subcommittee," "Eki-Hirobe Subcommittee," "Universal Design Team," and other groups were established. Strategy meetings, including debriefing sessions, have been held 14 times to date, and Odakyu Electric Railway, which had gradually changed from its initial attitude of refusing to engage in dialogue to one of positive dialogue, has formulated a development policy of "supportive development" that actively listens to the voices of citizens.

Regarding the specific spatial design of the upper Odakyu Line use, a major design firm and our laboratory collaborated to create a "design guideline" for private sites in 2014 regarding how buildings should be built. Here, we proposed a vision that utilizes the scale of the existing streetscape in the building design to create a space that makes people want to walk. As a result, small-scale buildings such as the "BONUS TRACK" were built on the land above the Odakyu Line. The 2.2km stretch of land above the Odakyu line is a continuous green space without boundaries between public and private land and is currently managed by a group of citizens called the Shimokita Horticulture Club. In addition, spaces for various purposes such as university satellite campuses, nursery schools, hot spring inns, student dormitories, shophouses, commercial facilities, and lodging facilities are lined up linearly, creating an interesting walking space that is unprecedented in Japan and has made a significant contribution to raising the quality of life in the surrounding residential environment.

The Meaning of Shimokitazawa's Urban Development for Today's Society

Shimokitazawa has two major meanings for Japanese society today. The first is its value as a case study of reusing disused urban infrastructure and realizing urban "green infrastructure" as a new environmental device. The Shimokitazawa district, which originally had few public squares and green spaces, lacked the capacity to consider biodiversity. It is hoped that the green chain above the Odakyu Line will bear fruit as an important green belt in Tokyo in the decades to come. The "green infrastructure" includes, broadly speaking, disaster prevention, resource recycling, landscape enhancement, and community maintenance, and the Shimokitazawa urban development example is expected to lead to the maturation of a green community mediated by greenery in the future.

The second and more important point is that, through various processes, citizens learned and practiced the process of participation and collaboration, overcoming political oppositional structures, and this was reflected in visible changes in the city. At the time of the initial conflict over the urban planning road, the city was in an uneasy situation of division due to the confrontation between the government and citizens; however, the change of the district mayor was a good opportunity for the government, citizens, and businesses to begin a dialogue on the same platform through continuous roundtable discussions and workshops. Citizens with strong "feelings" are actively involved in the management of greenery and plazas, indicating the potential for local self-governance by willing stakeholders and that the groundwork for developing future area management has been firmly laid.

<table>
<tr><td>Chapter 3
5</td><td>兵庫県姫路駅前のまちづくりの実践
市民参加による駅前広場の整備とエリアマネジメント</td></tr>
</table>

概要

　人口53万人の中核都市の玄関に当たるJR鉄道駅前の整備事業で、市民の声を代弁する形で行政の既存計画を変更修正し、専門家ワークショップ、市民ワークショップを繰り返すことで丁寧に関係者間の合意形成を重ね、最終的に歴史的文脈や眺望景観に配慮したウォーカブルな駅前空間の創造に貢献した。

　近年来、欧米では公共事業に積極的に市民が関わることが奨行されているが、わが国においては、それが十分に実行されているとは言い難い。わが国における市民参画のプロセスは、インフラ整備や公共建築の現場では行われつつあるが、どちらかというと事業が起因するネガティブインパクトを予測するパブリックコメントにとどまり、創造的なプロセスに市民が積極的に連携して関われる事例は少ない。また、ユーザーでありクライアントである納税者に情報が開示されずに、限られた行政スタッフと専門家によって巨額な公共事業が決められることも通常に行われており、市民が主体となって公共事業の合意形成に参加している事例はあまり見られない。

　一方、地方における公共事業、特に中心市街地における公共空間のデザインを考えた場合に、大都市とは別の問題として、市民力を背景とし、その地域らしさがデザインに取り入れられた地域間競争を意識した丁寧なまちづくりの手法が求められている。特に複数の公共事業が並行して実施されるような場合、それらの複数プロジェクトを相互に連携させ、全体として都市景観の価値を高めるような調整行為も重要である。

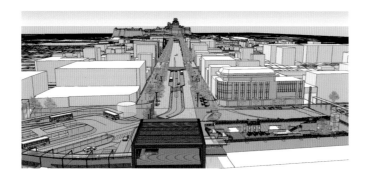

　兵庫県姫路市は、地方中核都市として連続立体交差事業などを機会に駅前再開発事業に取り組んできたが、2008年頃から公共空間デザインの合意形成に市民参加のプロセスを積極的に取り入れ、市民と専門家と行政が緊密に情報を交換しながら、計画や工事を進めるという新しい事業展開の手法を実施した。JRの新幹線と在来線、山陽電鉄、神姫バスなどが接続する交通結節点としての駅前広場であると同時に、世界遺産である姫路城への眺望をもつ環境広場としての歩行者空間

をいかに獲得するかという議論が大きな論点となった。さまざまなプロセスを経た結果、基本設計から竣工まで約4年という驚異的な短期間で、わが国ではまだ事例の少ないJR駅前の「トランジットモール」、地下街の再整備に合

（Photo1）再開発前の姫路駅前の狭隘な状況

わせた「サンクンガーデン」、市民が駅前で緑に触れあえる「芝生広場」、世界遺産である姫路城の新しい視点場となる「眺望デッキ」などが2015年までにほぼ実現した。2024年現在は完成した公共空間を今後どのように市民が使っていくかというエリアマネジメントの実装段階に入っている。ここでは、研究室が支援した事象を軸に、ここに至るまでに経緯の概略を紹介し、わが国が抱える公共空間のデザインに関する問題点を明らかにするとともに、今後の展望を考える。

駅前整備の背景と公共空間のデザインプロセス

　姫路市は、兵庫県西部の人口約53万人、面積約534km²を擁する中核市である。日本で初めて世界文化遺産に登録された姫路城を中心に、豊かな観光資源に恵まれ、多くの観光客が国内・海外から訪れている。市の都市核である姫路城〜JR姫路駅〜手柄山周辺エリアは古くから商業や交通の中心的役割を果たしてきたが、近年の社会経済情勢の変化に対応するため、姫路北駅前周辺整備計画が進められた。旧駅前広場は1957年に計画され、1959年に旧駅ビルと同時に完成したが、その後JRの地上駅舎・線路が南北交通を遮断したことによる交通問題が顕著となり、市中心部は踏切、迂回の高架橋の利用を余儀なくされた。1973年、市が高架化の基本構想を発表し、JR山陽本線等姫路駅付近連続立体交差事業、土地区画整理事業、関連道路事業などの面的・総合的開発が実施され2006年にはJR山陽本線の高架切替えが完了した。さらに市は都心部のまちづくりの指針となるグランドデザインとして「姫路市都心部まちづくり構想」を策定し、姫路城を望む北側駅前広場があるエントランスゾーンでは、旧駅ビルが移転した後のスペースを利用し、交通結節機能の向上・地下街の活性化などを可能にし、誰もが利用しやすい新しい駅前広場の計画に着手した。

市の基本レイアウトが変更され、市民案ができるプロセス

　姫路駅北駅前広場周りのデザインが実際に計画され、工事により完成する過程については、大まかに、（1）市の基本レイアウトが変更されるまでのプロセス、（2）基本設計・実施設計時における市民・行政・専門家の連携プロセス、（3）デザイン調整とエリアマネジメントプロセス、の3期に分けることができる。

第1期の展開プロセス

【各種団体による代替案】

　姫路市は、2006年3月に「姫路市都心部まちづくり構想」などを踏まえた整備計画の策定に取り組み、1987年に都市計画決定したプランを変更して、新北駅前広場について素案（Fig.1）を公表した。ここでは、地下街への排煙機能を兼ねたサンクンガーデンの設置や大面積のバスロータリー案が示されたが、これを契機に、駅前ロータリーの在り方に関心をもった市内の各種団体（商工会議所や商店街、市民など）から、市素案に対して複数の代替案が提案された。（Fig.2）

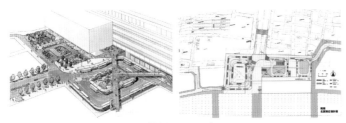

（Fig.1）市から提案された姫路駅北駅前広場のイメージ

（Fig.2）市内の各種団体から提案された代替案

【公募学生によるシャレットワークショップ】

　2008年から2009年にかけて、地元のNPO法人スローソサエティ協会とNPO法人まちづくりデザインサポートが中心となり、全国から集めた建築系学生による「学生による姫路シャレットワークショップ」を2回開催した。ここでは、ヒヤリングによる市民からの意見を取り入れ、姫路城を望む歩行者のための駅前空間のデザインが複数提案すると同時に文章化された10項目のアジェンダが発表された。模型や3Dシミュレーションなどによる発表内容はケーブルテレビや各種新聞にも

姫路の顔づくりを考える10の提言
1. 安全性と利便性を備えた新たな交通ジャンクションの創出
2. シンボルである姫路城への眺望の確保
3. 歩行空間の連携による回遊性の向上
4. 面的な広がりによる商業活動の振興
5. 多様なイベントが可能な広場空間の創出
6. 市民に親しまれる駅前空間の再構築
7. 既存地下街の再生による重層的な魅力空間の創出
8. 観光客や来街者に対するきめ細やかな配慮
9. 歴史を考慮した姫路らしい街並み景観の構成
10. 場の環境を形成する「緑」の適切な配置

（Fig.3）学生シャレットワークショップによる10の提言

掲載され、市民への周知の役割を果たした。またその後も一定期間成果を展示し、市民の意見をフィードバックしている。（Photo2）

（Photo2）学生によるシャレットワークショップ

【ラウンドテーブルとしての「市民フォーラム」】

　2009年4月にNPO法人スローソサエティ協会主催により、市長と駅前広場に係るバス会社や商業者代表などのステイクホルダーたちが一堂に顔を合わせる「市民フォーラム」が開催された。商業者がワークショップを経て作成した姫路市商店会連合会案も発表され、広場西側にバスとタクシーのロータリーの集約する計画案とトランジットモール（バス・タクシーのみを駅前に導入する計画）について、各団体の具体的な意見が明らかになり、今後の進展に大きな役割を果たした。（photo3,Fig.4）

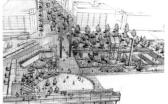

（Photo3）市民フォーラム　　　　（Fig.4）商店街連合会による提案
各団体の責任者が集合した　　　　駅前を歩行者優先にしている

【市から提示された3案と専門家ワークショップ】

　2009年6月になると姫路駅北駅前広場整備推進会議で、市から駅前広場のロータリーのレイアウトについて3案の選択肢が示された。それを受け、7月に交通計画や都市デザインの専門家を集めた「公開専門家ワークショップ」がNPO法人スローソサエティ協会主催により開催された。出席した専門家たちは市民の見守るなか、各案のメリット・デメリットを検証し、最終的に広場西側にバスとタクシーのロータリーの集約する「第3案」を推薦し、市長に提言を行った。その後、8月

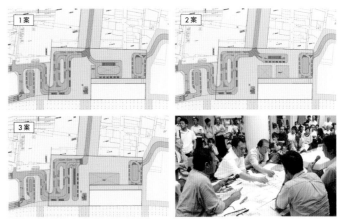

（Fig.5, Photo4）市による3案の提示と専門家ワークショップ

には市長が「第3案」で進めることを承認し、姫路駅北駅前広場整備推進会議において、でこの案を最終案とすることを正式に発表した。「第3案」には、姫路駅前にトランジットモールを整備すること、バス会社とタクシー会社による空間の共有、市民が安心して歩ける駅前広場の確保などの内容が含まれ、その後の駅前開発を決づける重要な方向付けが行われた。

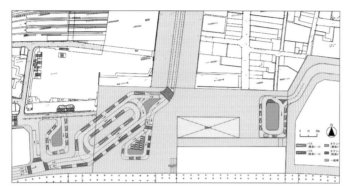

（Fig.6）市長が決定した最終案（第3案を基にしている）

第2期の展開

【基本設計・実施設計時における市民・行政・専門家の連携プロセス】

2009年度の基本設計および2010年度の実施設計と並行し、[連続セミナー][専門家会議][市民ワークショップ][推進会議]が有機的に開催され、セミナーによる意識化、専門家によるデザインの論理化 → 市民の検証と選択 → 関係者の合意形成、というプロセスが何回か繰り返され、大手前通りの歩車道配分、交番の配置、サンクンガーデンのデザイン方針、眺望デッキのデザイン方針、広場の活用法などのデザイン的課題が解決された。

姫路駅北駅前広場基本設計スケジュール

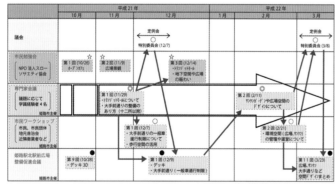

（Fig.7）有機的に展開された合意形成のフロー

【専門家の連続セミナーによる意識化】

NPO法人スローソサエティ協会主催により、「姫路の顔づくり」を勉強する連続セミナーが2009年秋に4回実施され、多様な立場の都市建築系専門家を招聘し、市職員と市民が一緒になって先行事例を中心に学ぶ機会を得た。会場を市役所の会議室としたことで、市の職員と

関心の高い市民が専門家の知識と経験を共有できたことはその後の合意形成に大きく貢献している。

【公開型の専門家会議によるデザインの論理化】

市の主催する「専門家会議」が3回開催された。市民たちに公開される形式で各種の専門家が模型を前に議論を展開し、妥当と思われるデザインを論理的に検証し、理由をつけて複数推薦する方向性を示

（Photo5）市民に公開された専門家会議

した。大手前通りにおける歩車道配分のパターン、サンクンガーデンのデザイン、広場のデザインなどが検討された。

私たちの研究室では、このサンクンガーデンのデザインをまとめるために、具体的な模型や3Dモデルを作成し、専門家会議の討議の材料を提供した。

【市民ワークショップによる検証と選択】

市の主催による「市民ワークショップ」が2回行われた。毎回、専門家会議おける検証内容の紹介と複数案についての論理的な解説の後に、グループに分かれて議論し、妥当と思われるデザインの方向性を選択

（Photo6）市主催による市民ワークショップ

し、公共空間の活用方法や管理などについての意見のぶつけ合いなどを行った。最終的には議論の内容を市民の意見として取りまとめ、毎回推進会議に伝えた。

【ステイクホルダーによる推進会議における承認】

市長を始め、駅前広場に関連する鉄道事業者、駅ビル、バス会社、商店街の代表者などが参加する姫路駅北駅前整備推進会議において、大手前通りの歩車道配分、交番の配置、サンクン

（Photo7）ステイクホルダーによる推進会議

ガーデンのデザイン方針、眺望デッキのデザイン方針、地上広場の使い方などについて、経緯が具体的に報告され、議論と承認がなされた。

第3期の展開

【デザイン調整とエリアマネジメント活動】

大手前通りや駅前広場の実施設計、工事時期に入ると、複数事業のデザインや工事を同時に調整する機能が必要となり、専門家グループ

による「姫路駅北駅前広場整備等デザイン会議」が市長の諮問機関として組織された。専門家たちは定期的に現場に集合し、現場から発生するデザイン的課題を解決すると共に、調整する方向を逐一市長に報告し承認を得た。今回の包括的な公共空間のデザインに一貫性がみられるのは、この会議が果たした役割が大きい。

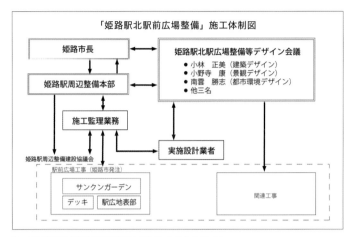

（Fig.8）姫路駅北駅前広場整備等デザイン会議の位置づけ

　一方、2009年ごろの市の基本レイアウトの変更から、完成後の活用・運営の課題や可能性を模索するための社会実験が行われた。
　2011年度には、実施設計に市民の意見を反映させるプロセスとして、行政、設計担当者、専門家と、将来広場の運営に大きく関わる事業者、地権者、市民団体などが集まり、姫路駅前広場活用準備会が数回開催された。これが市民にとって実施設計に直接意見を反映できる貴重な機会となった。ここでは、私たちの研究室は各会合の詳細な記録を取り、どのような組織や団体がどのような会議体で意思決定し、合意形成に至ったかを克明に記録した。
　この間の空間デザインのプロセスにおける合意形成の経緯をダイアグラム化したものを見ると、どのような会議体が関わりながら、全体の合意形成が図られていったかを理解することができる。まちづくりデザインにおいては、進行中のプロセスの中で行政・市民間の合意形成を図ることは非常に重要であるが、それに加え、それまでの経緯を分かりやすいダイアグラムにして「見える化」することで、一般市民の理解を得やすくし、公共的事業を進めることは極めて重要である（Fig.12）。

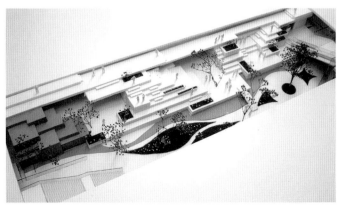

（Photo8）研究室で作成した模型によるシミュレーション
一般市民には最も理解されやすいツールである

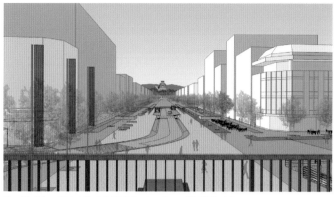

（Fig.9）スケッチアップによる3Dシミュレーション（眺望デッキ）

（Fig.10）スケッチアップによる3Dシミュレーション（サンクンガーデン）

　姫路北駅前周辺整備においては、駅ビルの移転に伴い、計画から竣工までの期間が極めて短いという厳しい条件のなかで、専門家が複数のデザイン案を示し、市民がそれを選択し、その後に関係者がそれを承認して行政と議会がそれを実行に移すというプロセスがシームレスに実行できたことは特筆すべき点である。当初は駅前の開発に対する市民の関心はあまり高いものではなかったが、計画と工事が進捗するにつれて、徐々に意識が高まり、現在ではエリアマネジメントが機能している。何回も実施されている社会実験ではサンクンガーデンを利用した結婚式やさまざまなイベントが企画され、単に物理的なSPACE（空間）であったものが、市民が関わることでPLACE（活動の場）に進化したことを確認できる。今回専門家や研究室が果たした役割は、情報の開示と透明化が合意形成のプロセスをスムーズにし、その後のトラブルを軽減させるという発見であった。一方で、ひとたび敷地を超えると職能的権限がなくなる建築基準法などの法制度が未だに都市デザインという複数の公共事業の調整機能を包含していないため、実際の現場での調整は極めて困難であった。複数の公共事業を繋ぎ合わせるための職能の不在は今後の大きな課題である。大学の研究室などの中立的組織によるボランティア活動に頼っているわが国の実状は、少しずつ変えていかなくてはならない。

（Fig.11）全体のマスタープラン
一般車のロータリーを東西に配置したので、メイン動線にはバスとタクシーだけが進入できるトランジットモールが実現した

（Photo11）地下商店街の上に展開された芝生広場

（Photo9,10）眺望デッキの外観と
内部からの眺望
時間をかけて姫路城を眺められる
視点場が生まれた

（Photo12）
サンクンガーデンの風景
姫路城で使われている、石・鉄・木に
建材を限定し、城と連携する
前庭として位置付けた

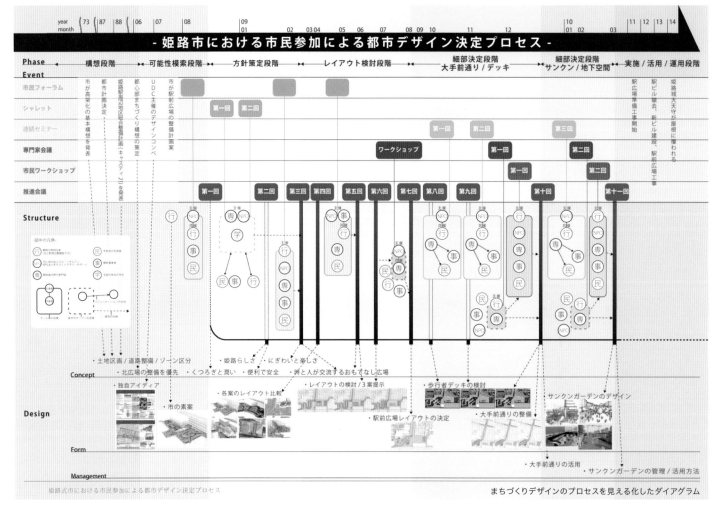

（Fig.12）まちづくりデザインのプロセスを見える化したダイアグラム

Development of the Plaza in front of the Station and Area Management through Citizen Participation

SUMMARY

In a project to develop the area in front of the JR railroad station, which serves as the entrance to a core city with a population of 530,000, the government's existing plan was modified in a manner that represented the voices of citizens. Through repeated expert workshops and citizen workshops, a consensus was carefully formed among the parties concerned, ultimately contributing to the creation of a walkable space in front of the station that takes into consideration its historical context and view of the landscape.

In recent years, active citizen involvement in public works has been encouraged in Europe and the United States. In Japan, however, such involvement has not yet been fully implemented. Citizen participation in the process of infrastructure development and public construction has been taking place, but it has been limited to public comments that predict the negative impact of projects, and there have been few cases in which citizens have been actively involved in the creative process. In addition, it is common practice for huge-cost public projects to be decided by a limited number of administrative staff and experts without disclosing information to taxpayers, who are the users and clients of the projects.

Public works projects in local cities, especially the design of public spaces in central city areas, require careful town planning methods that consider inter-regional competition, backed by citizen power and incorporating local characteristics in the design, which is a different issue from that in large cities. Especially in cases in which multiple public projects are to be implemented in parallel, it is important to coordinate actions that link these multiple projects and enhance the value of the urban landscape as a whole.

Himeji City, Hyogo Prefecture, is a core regional city that has conducted redevelopment projects in front of its station, taking advantage of opportunities such as a continuous multi-level crossing project. Since around 2008, the city has actively incorporated citizen participation in consensus building for public space design and implemented a new project development method in which citizens, experts, and the government closely exchange information during planning and construction. A major point of contention involved the design of a pedestrian space that would serve as an environmental plaza with a view of Himeji Castle, a World Heritage site, as well as a transportation node connecting the JR Shinkansen and conventional lines, Sanyo Electric Railway, and Shinki Buses. Over a notably short period of about four years from basic design to completion, the city constructed a "transit mall" in front of JR Station, of which there are still few examples in Japan; a "sunken garden" in conjunction with the redevelopment of the underground shopping center; a "grass plaza" in which citizens can come into contact with greenery in front of the station; and a "viewing deck" that provides a new perspective on Himeji Castle, a World Heritage site. As of 2024, the area is in the implementation stage of area management, in which citizens will decide how to use the completed public space. This section outlines the process that led to this point, focusing on the events supported by our laboratory, and clarifies issues related to the design of public spaces in Japan, as well as considering future prospects.

Background of Station Development and Design Process for Public Space
Himeji City is a core city in western Hyogo Prefecture with an area of 534 km2. Blessed with rich tourism resources centering around Himeji Castle, the city attracts many domestic and international tourists. The area around Himeji Castle, JR Himeji Station, and Mt. Tegara, the city's urban core, has long played a central role in commerce and transportation, but in response to recent changes in social and economic conditions, the Himeji North Station Area Development Plan has been promoted. The old station square was planned in 1957 and completed at the same time as the old station building in 1959, but later traffic problems became more pronounced due to JR's ground station building and tracks blocking north-south traffic, forcing the city center to use railroad crossings and bypass elevated bridges. In 1973, the city announced a basic plan for elevating the station, the JR Sanyo main line, and other lines. In 2006, the elevated JR Sanyo main line was completed. In addition, the city formulated the "Himeji City Center Town Development Concept" as a grand design to guide urban development in the city center, and in the entrance zone where the station square on the north side overlooking Himeji Castle is located, the space after the old station building was relocated to improve transportation functions, activate the underground shopping center, and to plan a new, more accessible station square.

The Process from Changing the City's Basic Layout to Citizen's Plan

The process through which the plaza in front of Himeji Station North Station was designed and constructed can be roughly divided into three phases: (1) changing the city's basic layout, (2) collaboration among citizens, government, and experts during the design and implementation phases, and (3) the design coordination and area management process.

The first phase is the development process

Alternatives by Various Organizations

In March 2006, Himeji City began working on a development plan based on the "Himeji City Center Urban Development Concept" and other plans, and released a rough draft (Figure 1) for the New North Station Square, changing the plan that had been approved in 1987. This led various groups in the city (including the chamber of commerce and industry, shopping district, and citizens) to propose several alternatives to the city's draft plan. (Fig. 2)

Charrette workshop by publicly recruited students

From 2008 to 2009, the local NPO Slow Society Association and NPO Machizukuri Design Support took the lead in holding two "Himeji Charrette Workshops by Students" with architecture-related students from across Japan. Here, an agenda of ten items was presented in writing as well as multiple proposals for the design of a space in front of the station for pedestrians overlooking Himeji Castle, incorporating opinions from the public through hearings. The presentations, which included models and 3D simulations, were broadcast on cable TV and in various newspapers and served to publicize the project. The results were also exhibited for a certain period of time to allow for the collection of public feedback. (Photo1)

Citizens' Forum as a Roundtable

In April 2009, the NPO Slow Society Association held a "Citizens' Forum" to bring together the mayor, representatives of bus companies and merchants, and other stakeholders in the station square. The forum also presented the Himeji City Federation of Merchants' Associations' proposal, which was developed through workshops held by the merchants, and clarified the specific opinions of each organization regarding the plan to consolidate the bus and cab rotaries on the west side of the square and the transit mall (a plan to introduce only buses and cabs in front of the station), which played an important role in the future development of the area. The first two plans are shown in Figure 3. (Photo3, Figure 4)

The three proposals presented by the city and the expert workshop

In June 2009, at the Himeji Station North Station Square Development Promotion Conference, the city presented three options for the layout of the roundabout in front of the station square. In response to this, an "Open Expert Workshop" was held in July, hosted by the NPO Slow Society Association, bringing together experts in transportation planning and urban design. The attending experts examined the merits and demerits of each proposal under citizens' watchful eyes and finally recommended "Proposal 3," which would consolidate the bus and cab roundabouts on the west side of the plaza, and presented it to the mayor. In August, the mayor approved the third proposal and officially announced it as the final proposal at the Himeji Station North Station Square Improvement Promotion Conference. "Proposal 3" included the construction of a transit mall in front of Himeji Station, the sharing of space between bus and cab companies, and the securing of a plaza in front of the station where citizens could walk in peace.
(Fig.5, Photo4)

Development of the second phase

Collaborative process among citizens, government, and experts during basic and implementation design phases

In parallel with the basic design in 2009 and the implementation design in 2010, a series of seminars, expert meetings, citizen workshops, and promotion meetings were held organically, and the process of raising awareness through seminars, logicalizing the design by experts, verification and selection by citizens, and consensus building by all parties involved was held several times. Design issues were resolved, such as the distribution of pedestrian and vehicular traffic on Otemae-dori, the placement of the police box, the design policy for the sunken garden, the design policy for the viewing deck, and the utilization of the plaza (Fig. 7).

Awareness raising through a series of seminars by experts

Four seminars to study the "face of Himeji" were held in the fall of 2009 under the auspices of the NPO Slow Society Association. Urban and architectural specialists from various positions were invited and city officials and citizens had the opportunity to learn together mainly from precedents. By using a conference room at City Hall as the venue, city staff and interested citizens were able to share the knowledge and experience of the experts, which greatly contributed to the subsequent consensus-building process.

Design Logic through Open Expert Meetings

The city hosted three "experts' meetings." Various experts gave presentations in front of models in a manner open to the public, logically verified designs that they considered appropriate, and gave reasons and directions for recommending multiple designs. Patterns for the distribution of pedestrian and vehicular traffic on Otemae Avenue, the design of the sunken garden, and the design of the plaza were all considered. Our laboratory created specific models and 3D models to present this sunken garden design and provided material for discussions at the expert meeting.

Verification and Selection through Citizen Workshops

Two "citizen workshops" were held under the auspices of the city. Each time, after an introduction of the verification contents of the expert meeting and logical explanations of multiple proposals, the participants were divided into groups to discuss, select a design direction that seemed appropriate, and exchange opinions regarding the utilization and management of the public space. Finally, the discussions were summarized as citizen opinions and conveyed to the promotion meeting. (Photo6)

Approval by stakeholders at the promotion meeting

At the Himeji Station North Station Improvement Promotion Conference, attended by the mayor, representatives of railroad companies, station buildings, bus companies, and shopping malls related to the station plaza, the process of the distribution of pedestrian and vehicular traffic on Otemae-dori, placement of police boxes, design policy for the sunken garden, design policy for the viewing deck, and use of the ground plaza were reported in detail for discussion and approval. (Photo7)

Phase 3 Development

Design coordination and area management activities

As the implementation design and construction of Otemae-dori Avenue and the station plaza began, it became necessary to coordinate the design and construction of multiple projects simultaneously. Thus, the "Himeji Station North Station Plaza Improvement Design Committee" was organized by a group of experts as an advisory body to the mayor. The experts met regularly at the site to resolve design issues and reported the direction of adjustment to the mayor for approval. This meeting played a significant role in the consistency of the comprehensive public space design.

However, since around 2009, when the city changed the basic layout, a major movement has emerged to conduct social experiments to explore the issues and possibilities of utilization and management after the completion of the project.
In FY2011, as a process to reflect citizens' opinions in the implementation design, a preparatory meeting for the utilization of Himeji Station Square was held several times, bringing together the administration, designers, experts, and business operators, landowners, and citizen groups that will be significantly involved in the future management of the plaza. This was a valuable opportunity for citizens to have their opinions directly reflected in the implementation design. Here, our laboratory took detailed records of each meeting, clearly documenting which organizations and groups made decisions and reached consensus.

A diagrammatic representation of how consensus was reached during the spatial design process provides an understanding of the types of bodies involved and the overall consensus that was reached. In urban development design, it is very important to build consensus among the government and citizens through an ongoing process. In addition, it is extremely important to visualize the process using easy-to-understand diagrams to facilitate the understanding of the general public and to promote public projects.((Fig.12)

In the development of the area in front of Himeji North Station, it is noteworthy that, under the severe conditions of the relocation of the station building and the extremely short period between planning and completion of construction, the process was seamless: the experts presented multiple design proposals, the citizens selected one, the relevant people approved it, and the administration and the council put it into action. Initially, citizens' interest in the development of the area in front of the station was not very high, but as planning and construction progressed, awareness gradually increased, and area management began to function. In the many social experiments that have been conducted, weddings and various events have been planned using the Sunken Garden, confirming that what was once simply a physical SPACE has evolved into a PLACE through the involvement of the citizens. The role played by experts and laboratories in this project was the discovery that disclosure and transparency of information facilitates the process of consensus building and reduces subsequent problems. However, the actual on-site coordination was extremely difficult because the legal system, such as the Building Standard Law, which loses its authority once a site is exceeded, still does not encompass the coordination function of multiple public projects in urban design. The absence of a professional function to link multiple public projects is a major challenge for the future. The actual situation in Japan, which relies on volunteer activities by neutral organizations such as university laboratories, must be gradually changed.

Photo 1　Narrow space in front of Himeji Station before redevelopment
Fig.1　Image of the plaza in front of Himeji Station North Station proposed by the city
Fig.2　Alternative plans proposed by various organizations in the city
Photo1　Charrette workshop by students
Fig.3　Ten recommendations from the student charette workshop
Photo2　Citizens' Forum
Fig.4　Proposal by the Federation of Merchants' Associations
　　　The responsible persons from each organization gathered in front of the station, giving priority to pedestrians
Fig.5, Photo 3　Presentation of three proposals by the city and expert workshop
Fig.6　Final plan decided by the mayor (based on the third plan)
Fig.6　Flow of consensus building developed organically
Photo 4　Expert meeting open to the public
Photo 5　City-sponsored citizen workshop
Photo6　Promotion meeting by stakeholders
Fig. 7　Positioning of the Himeji Station North Station Square Design Conference
Photo7　Simulation using a model
　　　This is the easiest tool for the general public to understand
Fig.8　3D simulation generated through sketch-up (viewing deck) (Fig.9) 3D simulation generated through sketch-up sunken garden)
Fig.10　Diagram visualizing the process of urban development design
Fig.11　Overall master plan
　　　The rotary for general traffic is located on the east-west side so that only buses and cabs can enter the main traffic line
Photo 8.9　Exterior and interior views of the viewing deck
　　　A viewpoint from which visitors can take their time to admire Himeji Castle was created
Photo 10　Grass plaza above the underground shopping mall
Photo 11　View of the Sunken Garden
　　　The Sunken Garden is positioned as a front garden in connection with the castle, using only stone, steel, and wood materials used in Himeji Castle

終 〜〜〜 おわりに
ENDNOTE

私には4人のメンターがいる。東京大学で直接師事した芦原義信先生、香山壽夫先生、建築設計の実務を教えていただいた丹下健三先生、ハーバード大学での指導教授ピーター・ロウ先生である。芦原先生からは、外部空間および街並みへの関心、建築家としての人間のあり方を学び、香山先生からは、設計における歴史の見方および設計の心を学んだ。丹下先生からは、戦略的なアーバンデザインの手法を具体的に学び、ロウ先生からは、実践的なアーバンデザインの研究と教育の方法を学んだ。また、学生時代から今まで、建築家槇文彦さんによる日本特有のアーバニズムと空間構成の理論からは常に強い影響を受け、東京大学の原弘司先生による数理的な根拠によるグラフ表現と視覚情報については、強い共感のもとに自分の研究に反映させてきた。今から振り返ると、これら先達から受けた知見や経験は、常に私自身の通底した思考基盤を支えており、脈々と研究・教育のエンジンとして息づいていたように思われる。これらすべてが大学という場所で発露されたということに気づかされる。

私が明治大学に赴任してからの前半20年間は、川崎生田キャンパスでさまざまな都市の「見える化」に関する研究実践を繰り返し、後半10年間は、中野キャンパスに新設された英語による大学院の国際建築都市デザインプログラム（IAUD）で、生田で得られた知見をアジアの都市研究へ応用することを試みはじめた。これについても少しずつ成果が表れ始めている。これらの長期の研究期間に共に研究に携わってくれた同僚の研究者たち、多くの学生たちに、格別の感謝の意を述べたい。

建築設計に関わる「運動論」について言えば、東京下北沢地区における20年にわたるまちづくりの展開、岡山県高梁市における30年間の継続的なまちづくりアクション、市民参加による姫路駅前の都市デザイン、さらに言えば、東京六本木の国際文化会館の逆転プロセスによる保存再生プロジェクトなど、どれをとっても単に敷地の中の建物の設計にはとどまらない、社会への働きかけ、あるいは異議申し立てのアクションの連続であったように思う。

わが国の大学における「敷地内」を中心とした建築教育のシステムを見直し、これからは敷地を超えたアーバンデザイン／アーキテクチャーの研究教育がさらに展開されることが望まれる。そのような想いもこめてこの本を出版することを思い立ったが、少しでも多くの人々や学生たちが、この本を機会に自分たちを取り巻く都市環境のありかたに関心を持っていただければ本望である。

まずは、この場を借りて、研究室のGIS研究を先導していただいた近畿大学の小池博先生、エリアマネージメント研究の領域を広げていただいた日本大学の泉山塁威先生に格別の感謝の意を述べたい。また、明治大学の同僚教員、日本建築学会、都市計画学会の同僚研究者たち、国際建築都市プログラムのパートナーとなったシンガポール国立大学のChye Kiang教授, Yunn Chii Wong教授、タイのチュラロンコン大学のPreechaya Sittipunt教授、Scot Drake教授、Antoine Lassus教授、米国ハーバード大学のMark Mulligan教授（交流当時）、米国カリフォルニア大学バークレー校のPeter, Bosselman教授、米国オレゴン大学のHoward Davis教授、Hans Neis教授、パリ国立建築大学ラヴィレッド校のMark Bourdier教授らとは頻繁に意見を交換し、アジアの建築や都市論についてさまざまな議論を行った。これらの知的交流も私達の研究の大きな支えとなっている。そして何よりも、研究やシャレットワークショップを共に経験した300名以上の小林ゼミナールの卒業生たちが国の内外で活躍していることは、教員にとって至福のプレゼントである。さらに彼らには応援のエールを送りたい。

最後に、さまざまな編集作業を支援してくれた和歌山公博くん、田村順子先生、無理な要求に最後まで対応していただいたエディトリアルデザイナーの横山ちなみさん、編集担当者の中川沙織さんに心から感謝の意を述べたい。

2024年3月　東京にて

I would like to acknowledge four of my mentors: Prof. Yoshinobu Ashihara and Prof. Hisao Kayama, under whom I studied at the University of Tokyo; Prof. Kenzo Tange, who taught me the practice of architectural design; and Peter Rowe, my advisor at Harvard University. Thanks to Prof. Ashihara, I developed an academic interest in exterior spaces and townscapes and learned how to be an architect as human being. From Prof. Kohyama, I learned how to see history through architectural design and the heart of design thinking. From Prof. Tange, I learned the specifics of strategic urban design methods, and from Prof. Rowe, I learned practical urban design research and teaching methods. From my student days until now, I have always been strongly influenced by architect Fumihiko Maki's theories on urbanism and spatial organization inherent to Japan, and have reflected the mathematical-based graphical representation and visual information of Prof. Hiroshi Hara of the University of Tokyo in my research with strong sympathy.

Looking back, the knowledge and experiences I gained from these predecessors have always supported my foundation of thought, and seem to have been the engine of my research and education. Now, I realize that all of this has stemmed from the broad network around the universities.

During the first twenty years of my appointment at Meiji University, I continuously conducted research on various urban "visualization" practices at the Kawasaki Ikuta Campus, and during the latter ten years, I began to apply the knowledge I gained at Ikuta to urban research in Asia at the newly established International Program in Architecture and Urban Design (IAUD) at the Nakano Campus, a graduate program in English. This program is also gradually beginning to show results. I would like to express my special thanks to my university colleagues and the many students who have worked with me during these long periods of research.

My "activism" in architectural design work has included the twenty-year development of the Shimo-Kitazawa district in Tokyo, the thirty-year continuous community development action in Takahashi City, Okayama Prefecture, the urban design in front of Himeji Station with citizen's participation, and the preservation and revitalization of the International House of Japan in Roppongi, Tokyo, through a reverse process, to name a few. In all of these projects, the design was not limited to simply thinking about a building on a site, but involved a series of actions that lobbied society or raised objections.
Architectural education at universities in Japan was established in the Meiji era (1868-1912) as the Zouke Gakka (Department of House Building) at the University of Tokyo, one of the educational goals of which was to provide students with the skills to construct safe and solid housing on a site. However, about 150 years have passed since then, and just as in those days, architectural education at universities still focuses on specifications and the design of buildings "on site" in accordance with the Building Standard Law. It is inevitable that research and education on urban design and architecture beyond the site will be further developed in the future. With this in mind, I have decided to publish this book and hope that as many people and students as possible will use it as an opportunity to become interested in the nature of the urban environment that surrounds them.

First, I would like to take this opportunity to express my special thanks to Prof. Hiroshi Koike of Kinki University for leading the GIS research in our laboratory, and Assoc. Prof. Rui Izumiyama of Nihon University for expanding the area management research. I would also like to thank my faculty colleagues at Meiji University, the Architectural Institute of Japan, and the City Planning Institute of Japan, as well as my partners in the International Program in Architecture and Urban Design at Meiji University; Prof. Chye Kiang and Yunn Chii Wong at the National University of Singapore; and Prof. Preechaya Sittipunt, Prof. Scot Drake, and Prof. Antoine Lassus at Chulalongkorn University in Thailand; Prof. Mark Mulligan of Harvard University, USA (at the time of the workshops); Prof. Peter Bosselman at the University of California, Berkeley, USA; Prof. Howard Davis and Prof. Hans Neis at the University of Oregon, USA; Prof. Mark Bourdier at École Nationale Supérieure d'Architecture, Paris Lavillette; and others, with whom I frequently exchanged views and engaged in various discussions on architecture and urbanism in Asia. These intellectual exchanges have also been a great support for our research. Above all, it is a blissful gift for me as a teacher that more than 300 graduates of the Kobayashi Seminar (Koba-ken), with whom I shared research and charrette workshops, are now active in Japan and abroad. Furthermore, I would like to send them my largest cheers of support.
Finally, I would like to express my sincere gratitude to Takahiro Wakayama and Assoc. Prof. Junko Tamura for their assistance in various editorial tasks, to Chinami Yokoyama, the editorial designer, and to Saori Nakagawa, the editor in charge, for their last-minute responses to my unreasonable requests.

March 2024, Tokyo

明治大学理工学部建築学科 都市建築デザイン（小林正美）研究室 卒業生リスト

期生	卒業生
1期生（1992）	池村圭造、小川俊之、榊法明、新関謙一郎、福嶋健一、元永二朗
2期生（1993）	嘉屋本准、桑原敦、佐藤仁昭、澤田浩一、神藤臣元、菅和禎、関戸拓司、新野裕之、馬場宏典、船越章央、前田道雄、矢沢真俊、山口英則、吉松猛、山本英広
3期生（1994）	池田秀幸、伊藤由理、唐戸禎光、木村久司、栗原秀志、栗屋健太郎、白石博昭、中島裕美子(旧姓中込)、長友寛昌、中津留章仁、西久保毅人、野中智康、秦将之、平野明美、古田昌之、前田英則、増山士郎、三木隆靖、山下裕加子、吉村昌也
4期生（1995）	飯田伸司、石井和英、斎宮隆行、岩本圭右、浦山真吾、北見啓輔、炭谷麻子、野神進、野島伸一、畑川拓二、蒔田佳代、松本祥忍
5期生（1996）	相川隆弘、石毛厚史、太田繁夫、小杉学、佐伯聡子、標由理、白木直子、新谷織恵、滝口聡司、竹地直記、西田直哉、古市修、宝神尚史
6期生（1997）	安孫子崇弘、池田誠、片山剛巨、神谷園子、草間詠子、小山貴臣、佐伯恵理子、阪本泰智、嶋田泰子、高瀬みどり、高橋顕也、田中喜仁、中塚健仁、西村周作、宮川大介、三宅良一、宮中芳樹、山内大輔
7期生（1998）	伊藤成人、岡野大介、柏木尚浩、加藤政宏、川久保哲二、瀬端江美子、高崎由紀子、津村彰、野呂竜彦、三宅陽介
8期生（1999）	安藤寿洋、今井達哉、大垣知美、大嶽嘉数也、大橋佳子、片山千都世、小池要、寺本美千代、中島鶴人、野村昭、畑本昌輝、布施昇、前田卓、水野明洋
9期生（2000）	青木康明、有冨智也、板屋孝治、池田彩子、井上愛之、片山研、加藤花菜、加藤慶介、神谷桂太、神谷優治、小島隆司、津久井祐介、野田晋一郎、飛高達也、宮廻景子、村上浩平、森友峰
10期生（2001）	田貴史、榎戸孝行、大久保洋、岡晋輔、唐鎌豊、角田孝一、中路智子、平手健一、藤森亮、堀内功太郎、本間　桂、松谷浩平、村上祐資、八木幸助、山添直樹、横井隆、渡辺朋恵
11期生（2002）	青羽研一郎、磯村歩未、伊東潤、小川貴之、乙武正宏、北村伸二、牛腸将史、高塚陽介、高橋智香、陳敦琳、坪井和子、町田怜子、森啓将、森下千晶、山崎寛
12期生（2003）	秋山千尋、井川雅裕、李康熙、植田博文、大向聡、小野健二、加藤智里、河野修、興水真利子、斎藤仁、正田和之、辻康宏、平井孝広、森本通生、山中進吾
13期生（2004）	伊藤光、吉津谷知美、高橋秀樹、竹内一真、釣佳彦、德澤真希、中村浩樹、パーソンシュク・センターノンバット、布施大輔、洞口和也、横山正一、吉田達乃鯉
14期生（2005）	安藤哲也、鵜飼高生、宇賀神幸、栗田梓、佐々木翔、首藤涼兵、新宮あやこ、高谷俊介、田中みぶ奈、豊木ユウサン／エレン、西真紀子、八木厚輔、山本悟史
15期生（2006）	荒巻菜生子、加藤健介、姜性湖、坂本一樹、佐々木裕也、佐藤秀光、柴広朗、鈴木義大、宗木綾子、森本絢子、山川智嗣
16期生（2007）	池田翔太、今井麻穂、岡村咲子、西條公晴、竹内智、竹鼻慧一、中村敬一、東原大輔、藤田健児、武藤夏香、持丸卓也
17期生（2008）	岩寺静香、岸上和樹、黒田美知子、小早川武朗、高橋真由美、平川勇登、福田浩士、水野克哉
18期生（2009）	内野琢麿、川島悠都、監物契、谷川将光、富樫広貴、羽根田誠、丸山洋平、村中奈々
19期生（2010）	秋山遥真、秋山弘樹、阿部大輔、古跡匡、曽根高麻世、高崎翔太、田邊剛士、中川沙織、仁藤有理、武藤雅昭
20期生（2011）	稲田悠輔、小原えり、加藤夕佳、加藤遼、笹木健太、新崎希美、萩野日向子、山口玲、山本真兵
21期生（2012）	泉山塁威、荻野航、小林佑輔、鈴木脩斗、長谷川祥、秦野早紀
22期生（2013）	阿部 大和、木村麻優子、黒木美沙、重田力美、林田咲紀、藤佳紀、堀井皓生、松尾悠昂
23期生（2014）	秋山 祐毅、池田敬一郎、木村肇、後藤 雄二、田代祐一、泊 絢香、松崎 航大、茂木 祐介
24期生（2015）	石坂 亮典、江口 春花、神谷 大道、木曽 俊樹、林 愛弥、日野 将徳、南澤 隆広、山本 千尋
25期生（2016）	安藤輝、加藤悠輔、小菅歩、佐保田幸美、田辺 優里子、早坂覚啓、古川優衣
26期生（2017）	泉翔希、葛璇宇、川上勇一郎、SAMIEI ESPINOZA ATUSA KAORI、大藤遼太、永野渓、長町遥、宮本紘花、和歌山公博
27期生（2018）	伊藤潤、小屋崚祐、曹雨、成田遼、向野岳、孟慶元、林奕君
28期生（2019）	飯山真生、上葛貴文、王嘉淳、國田佑亮、曹ホウ寧、丹波哲哉、陳丹芸、富沢颯太、冨谷竜司、水野綾子
29期生（2020）	新安萌音、洲崎志織、鈴木柚葉、詹洪玥梦、宋斯佳、丁文清、中山稚菜、西山慧、長谷川雅
30期生（2021）	秋本泰伸、北有希、佐藤颯、鈴木貴慈、高津拓海、張琳、長瀬智哉、山田彩月、WANG DANNI
31期生（2022）	有田昂平、王柳涵、加島聡之助、小池耀介、西方理奈、服部友香、馬新テイ、山本暁、吉村宇矢
32期生（1923）	石見優太、王逸函、塩崎末琴、須藤陽大、田畑駿、服部祥之、廣田芽生、柳茉奈歩

主要研究者リスト

No.	タイトル	研究者
4	GISによる都市の見える化	小池博
1-1	交換テストによる街並みのシミュレーション	中路智子、村上浩平、曽根高麻世
2-1	差異面による街並み形態のシミュレーション	竹地直記
2-2	差異面による景観色差のシミュレーション	神谷優治
3-1	光源投射法による街路空間のシークエンス	安藤寿洋
3-2	光源投射法による建物の透過性と空間シークエンス	本間桂
4-1	点群による「見る」室内空間のシークエンス	中山稚菜、高津巧海 ほか
4-2	点群による「見る」街路空間のシークエンス	長谷川雅、長瀬智也 ほか
4-3	点群による室内の色彩分析	冨谷竜司 ほか
5-1	下北沢（東京都）におけるユビキタス実験	辻康宏
5-2	高梁市（岡山県）におけるユビキタス実験	加藤健介
	建物用途の「雑多度」と地域の特徴	津久井祐介、和歌山公博
7-1	建物類型の集積度の解析	古市修
72	都市のグレインの視覚化	高橋顕也
8-1	公開空地の利用度の解析	片山千都世
8-2	東京都における公開空地の集積度分析	森友峰
9	東京都における超高層開発の分布予測	八木幸助
10	市民参加とエリアマネジメント	泉山塁威
1	登戸土地区画整理のためのデザインガイドライン	古市修
2	横須賀市における眺望景観ガイドライン	竹内一真

研究室が発表した関連論文

日本建築学会技術報告集

『「まちづくり」における「シャレットワークショップ」の
実験と評価に関する研究 岡山県高梁市における継続的ケーススタディー』
著：小林正美, 古市修 日本建築学会技術報告集
第15号 p283-288 2002年6月
"RESEARCH ON THE EXPERIMENT AND THE FEEDBACK OF "CHARRETT WORKSHOP" IN THE TOWN BUILDING PROCESS
-A continuous case studies in the city of Takahashi, Okayama-",
Masami Kobayashi, Osamu Fruichi, AIJ J. Technol Des. No.15, p283-288, Jun.2002
https://www.jstage.jst.go.jp/article/aijt/8/15/8_KJ00004057211/_article/-char/ja

『環境デザインにおけるアーバンデザイン教育の再評価に関する研究：
ハーバード大学大学院との同時並行演習（スタジオ）の事例比較を通じて』
著：小林正美 日本建築学会技術報告集
第16号 p343-348 2002年12月
"REVIEW OF URBAN DESIGN EDUCATION IN GRADUATE SCHOOL AS ENVIRONMENTAL DESIGN:
Comparative study of the urban design studios concurrently executed among Harvard University and Japanese Universities",
Masami Kobayashi, AIJ J. Technol Des. No.16, p343-348, Dec.2002
https://www.jstage.jst.go.jp/article/aijt/8/16/8_KJ00004655537/_article/-char/ja

『大学による継続的な町づくり支援の方法に関する研究
登戸区画整理事業における専門的支援』
著：古市修, 小林正美, 田中友章 日本建築学会技術報告集
第24号 p387-392 2006年12月
"STUDY ON THE ASSISTANCE BY UNIVERSITY TO LOCAL TOWN MAKING -Professional assistance in the land readjustment
project in Noborito district-",
Osamu Fruichi, Masami Kobayashi, Tomoaki Tanaka, AIJ J. Technol Des. No.24, p387-392, Dec.2006
https://www.jstage.jst.go.jp/article/aijt/12/24/12_KJ00004439681/_article/-char/ja

『ユビキタス技術を応用した街なかの回遊性の実験と評価に関する研究
小田急線「下北沢駅」周辺商店街におけるケーススタディ』
著：小林正美, 辻康宏, 元永二朗 日本建築学会技術報告集
第24号 p407-410 2006年12月
"EXPERIMENT AND EVALUATION OF PEDESTRIANS'CIRCULATION IN TOWN, APPLYING UBIQUITOUS TECHNOLOGY
-Case Study in the District of Shimokitazawa Station of Odakyu Line-",
Masami Kobayashi, Yasuhiro Tsuji, Jirou Motonaga, AIJ J. Technol Des. No.24, p407-410, Dec.2006
https://www.jstage.jst.go.jp/article/aijt/12/24/12_KJ00004439685/_article/-char/ja

『実践教育としてのまちづくりシャレットワークショップの研究
―学生参加のシャレットワークショップを事例として―』
著：高橋潤, 小林剛士, 小林正 日本建築学会技術報告集
第16巻 第33号 p711-716 2010年6月
"RESEARCH OF URBAN DESIGN CHARRETTE WORKSHOP AS PRACTICAL EDUCATION
-A case study of charrette workshop of student participation-",
Takahachi, Takeshi Kobayashi, Masami Kobayashi, AIJ J. Technol Des. Vol.16, No.33, p711-716, Dec.2006
https://www.jstage.jst.go.jp/article/aijt/16/33/16_33_711/_article/-char/ja

『市民参加型まちづくりにおけるユビキタス技術の導入に関する研究
―大山街道アクションフォーラムの取組みを通じた
常設ユビキタスシステムの実装―』
著：泉山塁威, 加藤健介, 小林正美, 小池博 日本建築学会技術報告集
第18巻 第39号 p727-732 2012年6月
"A STUDY ON THE INTRODUCTION OF UBIQUITOUS TECHNOLOGY IN TOWN PLANNING OF CITIZEN PARTICIOATION
-Implementation of a permanent UBIQUITOUS system through the projects of action forum of "OOYAMA-KAIDO"-",
Kensuke Kato, Masami Kobayashi, Hiroshi Koike, AIJ J. Technol Des. Vol.18, No.39, p727-732, Jun.2012
https://www.jstage.jst.go.jp/article/aijt/18/39/18_727/_article/-char/ja

『場所価値評価法を用いた景観評価に関する研究
― 東京・台東区浅草寺周辺の場所の評価 ―』
著：曽根高麻世, 小林正美 日本建築学会技術報告集
第26巻 第63号 p701-706, 2020年6月
"STUDY ON LANDSCAPE EVALUATION USING LOCATION EVALUATION METHOD
-Evaluation of Senso-ji temple neighborhood, Taito ward, Tokyo- ",
Mayo Sonetaka, Masami Kobayashi, AIJ J. Technol Des. Vol.26, No.63, p701-706, Jun.2020
https://www.jstage.jst.go.jp/article/aijt/26/63/26_701/_article/-char/ja

『パリの18-19世紀高級住宅における回遊動線と
可視領域のヴィジュアル化
―スペースシンタックス理論による平面解析のケーススタディ―』
著：酒井映命, 小林正美 日本建築学会技術報告集
第28巻 第69号 p858-863 2022年6月
"VISUALIZATION OF THE CIRCULATION AND THE VISIBLE AREA OF HIGH-GRADE APARTMENTS IN PARIS IN THE ERA OF 18-19TH
CENTURY -Case studies of plan analysis with Space Syntax theory-",
ri Sakai, Masami Kobayashi, AIJ J. Technol Des. Vol.28, No69, p858-863, Jun.2022
https://www.jstage.jst.go.jp/article/aijt/28/69/28_858/_article/-char/ja

『地域の空間的特徴に関する研究 ～「着彩用途複合ダイアグラム」と
「道路立体グラフ」を用いた定量的分析の試み～』
著：和歌山公博, 小林正美 日本建築学会技術報告集
第29巻 第73号 p1659-1664 2023年10月
"RESEARCH ON THE SPATIAL FEATURE OF NEIGHBORHOOD
~An attempt of quantitative analysis using the "Colored composite use diagram" and the "Road three-dimensional graph"~",
Takahiro Wakayama, Masami Kobayashi, AIJ J. Technol Des. Vol.29, No.73, p1659-1664, Oct.2023
https://www.jstage.jst.go.jp/article/aijt/29/73/29_1659/_article/-char/ja

日本建築学会計画論文集

『市街地再生手法における目標空間イメージ支援ツールの研究：
墨田区K地区における商店街再生計画のケーススタディ』
著：関谷浩史, 岡井敦, 小林正美 日本建築学会計画論文集
第559号 p145-152 2002年9月
"A STUDY ON GOAL IMAGE SUPPORT TOOL FOR THE METHOD OF A CITY REPRODUCTION
-A case study of the shopping district reproduction planning in Sumida-ku, K-area-",
Hiroshi Sekiya, Atsushi Okai, Masami Kobayashi, Journal of Architecture Planning, AIJ, No.59, p145-152, Sep.2002
https://www.jstage.jst.go.jp/article/aija/67/559/67_KJ00004075645/_article/-char/ja

『市街地再生手法における目標空間イメージ支援ツールの研究（その2）
WEB端末を活用した商店街再生計画案策定のケーススタディ』
著：関谷浩史, 岡井敦, 小林正美 日本建築学会計画論文集
第576号 p37-44 2004年2月
"A STUDY ON GOAL IMAGE SUPPORT TOOL FOR THE METHOD OF A CITY REPRODUCTION PART II
-A case study of the shopping district reproduction by use of personal computer-",
Hiroshi Sekiya, Atsushi Okai, Masami Kobayashi, Journal of Architecture Planning, AIJ, No.576, p33-44, Feb.2004
https://www.jstage.jst.go.jp/article/aija/69/576/69_KJ00004227008/_article/-char/ja

『街並み景観データベースを活用した歴史的街並み再生の
方法論に関する研究 ―岡山県高梁市における景観構造の視覚化と
町並み助成制度による修景効果の検証―』
著：古市修, 小林正美, 泉山塁威, 野口弘行, 内山善明、日本建築学会計画論文集
第77巻 第673号 p619-628 2012年3月
"A STUDY ON THE METHODLGY FOR REGENERATION OF HISTORICAL TOWNSCAPE UTILIZUBG URBAN DATABASE
-Test of the effectiveness of subsidy Program for townscapes in the city of Takahashi pref. Okayama-",
Osamu Fruichi, Masami Kobayashi, Rui Izumiyama, Hiroyoshi Noguchi, Yoshiaki Uchiyama, Journal of Architecture Planning,
AIJ, Vol.77, No.673, p619-628, Mar.2012
https://www.jstage.jst.go.jp/article/aija/77/673/77_619/_article/-char/ja

『都心部における「民有地の公共空間」の活用マネジメントに関する研究
―「東京のしゃれた街並みづくり推進条例」・まちづくり団体登録制度の
調査・分析を通して―』
著：泉山塁威, 加藤健介, 小林正美, 小池博 日本建築学会技術報告集
第80巻 第710号 p915-922 2015年4月
"STUDY ON THE APPLICATION AND MANAGEMENT OF PRIVATELY OWNED PUBLIC SPACES" IN THE URBAN CENTRAL AREA —
Through research and analysis to community management organization registration system the
"Tokyo Municipal Ordinance on promoting the creation SYARETA-MACHINAMI of Tokyo —",
Izumiyama, Hiroki Akiyama, Masami Kobayashi, Journal of Architecture Planning, AIJ, Vol.80, No.710, p915-922, Apr.2015
https://www.jstage.jst.go.jp/article/aija/80/710/80_915/_article/-char/ja

PROFILE

小林 正美
Masami Kobayashi

明治大学理工学部教授 工学博士
建築家／都市デザイナー

1954年東京生まれ。東京大学大学院修士課程修了の後、丹下健三・都市建築設計研究所勤務。フルブライト奨学金にて米国留学し、ハーバード大学大学院修士課程修了。帰国後、東京大学大学院博士課程修了。明治大学理工学部専任講師、助教授を経て現職。
ハーバード大学客員教授（2002）
カリフォルニア大学バークレイ校客員研究員（2007）
NPO法人「まちづくりデザインサポート」理事長
アルキメディア設計研究所 主宰
東京都台東区景観審議会会長、川崎市横須賀市景観審議会会長

専門は建築設計および都市デザイン論。「シャレットワークショップ」の手法により全国各地でまちづくり活動に参加。東京の下北沢地区、岡山県高梁市、兵庫県姫路市などの都市デザインを具体的に手掛ける。明治大学では、本邦初の完全英語による建築教育プログラム（IAUD）を開設し、国際的に活躍する専門家を育成している。

Architect/Urban Designer
Professor of Meiji University, Doctor of Engineering

1954 Born in Tokyo/ 1977 Bachelor of Architecture, 1979 Master of Architecture, University of Tokyo/ 1979-1986 Kenzo Tange & Associates/1988 Master of Design Studies, Graduate School of Design, Harvard University (with Fulbright Scholarship)/ 1989 PhD., University of Tokyo/2002 Visiting Professor of Harvard University / 2003-Professor of Meiji University /2007 Visiting Scholar of University of California, Berkeley

主要著書

「市民が関わるパブリックスペースデザイン」（エクスナレッジ2015）
「シモキタらしさのDNA」（エクスナレッジ2015）
「歴史的町並み再生のデザイン手法」（エクスナレッジ2013）
「インターベンションII（都市への介入）」（鹿島出版会2003）
「東京再生」（学芸出版　共著2003）
「ボストン建築探訪」（丸善1991）

Public Space Design with Citizen's Participation (X-Knowledge 2015)
DNA of Shimokitazawa(X-Knowledge 2015)
Regeneration of Historical Assets through "Charette Workshop" (X-Knowledge 2013)
Intervention II (Kajima Press 2003)
Regeneration of Tokyo (Gakugei Shuppan, co-author 2003)
Boston Architectural Exploration (Maruzen 1991)

受賞歴

2023年 アジア都市景観賞（高梁市のまちづくり）
2019年 文部科学大臣表彰科学技術賞
2018年 日本建築学会賞（業績部門）
　　　　「学生設計優秀作品展・建築・都市・環境」による建築設計教育への貢献
2016年 グッドデザイン特別賞（地域づくりデザイン賞）
　　　　「姫路駅北駅前広場および大手前通り」プロジェクト
2015年 日本建築学会教育賞
　　　　『岡山県高梁市における「シャレットワークショップ」手法による大学連携まちづくり教育への継続的取り組み』
2008年 土木学会デザイン賞最優秀賞受賞
　　　　「学びの森」プロジェクト
2007年 日本建築学会賞（業績部門）受賞
　　　　「国際文化会館の保存再生」

2023 Asian Urban Landscape Award (Takahashi City Planning)
2019 The Commendation for Science and Technology by the Minister of Education, Culture, Sports, Science and Technology
2018 Architectural Institute of Japan Award (Achievement Category)
Contribution to architectural design education through "Excellent Student Design Works Exhibition, Architecture, City and Environment"
2016 Good Design Special Award (Community Development Design Award)
Himeji Station North Station Square and Otemae Street Project
2015 Architectural Institute of Japan Education Award
Continuous efforts for education on town building in Takahashi City, Okayama Prefecture, through university collaboration using the "Charrette Workshop" method.
2008 Japan Society of Civil Engineers Design Award, Grand Prize
Forest of Learning" Project
2007 Architectural Institute of Japan Award (Achievement Category)
Conservation and Restoration of the International House of Japan

都市の「見える化」でまちが変わる
Urban Visualization Changes our Town

2024年3月19日　初版第1刷発行

著者： 小林正美＋明治大学都市建築デザイン研究室
Author: Masami Kobayashi + Meiji University Urban & Architectural Design Laboratory

発行者： 三輪浩之
Publisher: Hiroyuki Miwa

発行所： 株式会社エクスナレッジ
Publish Office: X-Knowledge Inc.

〒106-0032 東京都港区六本木7-2-26
https://www.xknowledge.co.jp/

編集 Tel. 03-3403-1343　Fax. 03-3403-1828
　　info@xknowledge.co.jp

販売 Tel. 03-3403-1321　Fax. 03-3403-1829